TECHNICAL REPORT

Planning Tool to Support Louisiana's Decisionmaking on Coastal Protection and Restoration

Technical Description

David G. Groves • Christopher Sharon • Debra Knopman

Sponsored by the Coastal Protection and Restoration Authority of Louisiana

RAND **GULF STATES POLICY INSTITUTE**

A study by RAND Infrastructure, Safety, and Environment

This research was sponsored by the Coastal Protection and Restoration Authority of the State of Louisiana and was conducted in the RAND Gulf States Policy Institute and the Environment, Energy, and Economic Development Program within RAND Infrastructure, Safety, and Environment.

Library of Congress Control Number: 2012947921

ISBN: 978-0-8330-7698-4

Published 2012 by the RAND Corporation
1776 Main Street, P.O. Box 2138, Santa Monica, CA 90407-2138
1200 South Hayes Street, Arlington, VA 22202-5050
4570 Fifth Avenue, Suite 600, Pittsburgh, PA 15213-2665
RAND URL: http://www.rand.org/
To order RAND documents or to obtain additional information, contact
Distribution Services: Telephone: (310) 451-7002;
Fax: (310) 451-6915; Email: order@rand.org

Preface

Coastal Louisiana's built and natural environment faces risks from catastrophic tropical storms, such as Hurricanes Katrina and Rita in 2005 and Gustav and Ike in 2008. Hurricanes flood cities, towns, and farmlands, forcing evacuations, damaging and destroying buildings and infrastructure, eroding coastal habitats, and threatening the health and safety of residents. Concurrently, the region is experiencing a dramatic conversion of coastal land and associated habitats to open water and a loss of important services provided by such ecosystems. The State of Louisiana, through its Coastal Protection and Restoration Authority (CPRA), responded to the threat of catastrophic hurricanes and ongoing land loss by engaging in a detailed modeling, simulation, and analysis exercise, the results of which informed *Louisiana's Comprehensive Master Plan for a Sustainable Coast* (CPRA, 2012c).

The Master Plan defines a set of coastal risk-reduction and restoration projects to be implemented in the coming decades to reduce hurricane flood risk to coastal communities and restore the Louisiana coast. When selecting projects to reduce the flood effects of hurricanes, CPRA evaluated the extent to which each project might reduce damage. Similarly, when choosing projects to restore the landscape, CPRA evaluated the extent to which each project might sustain or build new land and support various ecosystem-service benefits to the region. Based on these evaluations, risk-reduction and restoration projects were selected to provide the greatest level of risk-reduction and land-building benefits under a given budget constraint while being consistent with other objectives and principles of the Master Plan.

CPRA asked RAND to support the development of the Master Plan. One RAND project team, with the guidance of CPRA and other members of the Master Plan Delivery Team, developed a computer-based decision-support tool, called the CPRA Planning Tool. The Planning Tool provided technical analysis that supported the development of the Master Plan through CPRA and community-based deliberations. The Master Plan was presented to the Louisiana legislature in April 2012 and adopted for approval on May 22, 2012. CPRA supported a Technical Advisory Committee (Planning Tool—TAC), made up of three national experts on coastal and natural resource planning, to provide technical review of the Planning Tool and this document. Another RAND team developed a new model of coastal hurricane flood risk to evaluate risk-reduction projects in support of the Master Plan, to be described in another RAND document (Fischbach et al., forthcoming).

This document seeks to provide an accessible technical description of the Planning Tool and associated analyses used to develop the Master Plan. The intended audience includes planners, stakeholders, and others in Louisiana and elsewhere in the United States and in other countries who are interested in understanding the technical basis for the investments proposed in the Master Plan.

The RAND Environment, Energy, and Economic Development Program

This research was conducted in the Environment, Energy, and Economic Development Program (EEED) within RAND Infrastructure, Safety, and Environment (ISE). The mission of ISE is to improve the development, operation, use, and protection of society's essential physical assets and natural resources and to enhance the related social assets of safety and security of individuals in transit and in their workplaces and communities. The EEED research portfolio addresses environmental quality and regulation, energy resources and systems, water resources and systems, climate, natural hazards and disasters, and economic development—both domestically and internationally. EEED research is conducted for government, foundations, and the private sector.

Information about EEED is available online (http://www.rand.org/ise/environ). Inquiries about EEED projects should be sent to the following address:

Keith Crane, Director
Environment, Energy, and Economic Development Program, ISE
RAND Corporation
1200 South Hayes Street
Arlington, VA 22202-5050
703-413-1100, x5520
Keith_Crane@rand.org

RAND Gulf States Policy Institute

RAND created the Gulf States Policy Institute in 2005 to support hurricane recovery and long-term economic development in Louisiana, Mississippi, and Alabama. Today, RAND Gulf States provides objective analysis to federal, state, and local leaders in support of evidence-based policymaking and the well-being of individuals throughout the Gulf Coast region. With offices in New Orleans, Louisiana, and Jackson, Mississippi, RAND Gulf States is dedicated to helping the region address a wide range of challenges that include coastal risk reduction and restoration, health care, and workforce development. More information about RAND Gulf States can be found at http://www.rand.org/gulf-states/.

Questions or comments about this report should be sent to the project leaders, David Groves (David_Groves@rand.org) or Debra Knopman (Debra_Knopman@rand.org).

Contents

Figures

Tables

Summary

Louisiana's Coastal Crisis

Coastal Louisiana is on an unsustainable trajectory of ongoing conversion of coastal land to open water and increasing hurricane flood risk. Since the 1930s, 1,800 square miles of land have been lost to open water (Couvillion et al., 2011). This loss of land is changing the nature of the coastal environment profoundly and diminishing many of its benefits, including habitats for commercially and recreationally important species. Land loss is also decreasing the region's natural buffer against hurricane storm surges.

The causes of the ongoing land loss are varied and include natural and human-caused land subsidence, rising sea level, and the loss of nourishing sediment from Mississippi river flows that is now deposited deep in the Gulf of Mexico. Without major investments in coastal restoration, the Coastal Protection and Restoration Authority (CPRA) estimates that an additional 800 square miles could be lost over the next 50 years under moderate assumptions about future conditions, and 1,800 square miles under less optimistic assumptions (CPRA, 2012a). As communities and economic assets grow during the coming decades, the land that provides a protected buffer against storm surges is anticipated to continue to degrade. Sea-level rise and subsidence rates may accelerate (Vermeer and Rahmstorf, 2009; Kolker, Allison, and Hameed, 2011), and hurricanes may increase in frequency and magnitude in response to changing climate patterns (Knutson et al., 2010). As a consequence, flood risk is expected to rise significantly if further investments in risk-reduction and restoration projects are not made.

The Louisiana Comprehensive Master Plan and Planning Tool

To address this challenge, CPRA developed *Louisiana's Comprehensive Master Plan for a Sustainable Coast* (CPRA, 2012c), a 50-year plan for reducing hurricane flood risk and achieving a sustainable landscape. As part of this effort, CPRA supported the development of a computer-based decision-support tool called the *Planning Tool*. The Planning Tool was designed to support a *deliberation-with-analysis process* by which quantitative analysis is used not to provide a single answer but rather to frame and illuminate key policy trade-offs (National Research Council, 2009). Specifically, the Planning Tool helped CPRA to (1) make analytical and objective comparisons of hundreds of different risk-reduction and restoration projects, (2) identify and assess groups of projects (called *alternatives*) that could make up a comprehensive solution, and (3) display the trade-offs interactively to support iterative deliberation over alternatives.

Comparing Individual Risk-Reduction and Restoration Projects

The Planning Tool compares the ways in which individual projects affect the main objectives of the Master Plan—reducing hurricane flood risk and building and maintaining the coastal landscape. The Master Plan analyzed more than 40 structural risk-reduction projects, including levees and floodwalls, and nonstructural programs across the coast that reduce flood damage to residential and commercial structures through elevating, flood-proofing, or removing the structures. The Master Plan also analyzed approximately 250 restoration projects, including bank stabilization, barrier island restoration, channel realignment, sediment diversion, hydrologic restoration, marsh creation, oyster barrier reef, ridge restoration, and shoreline protection (Figure S.1).

The Planning Tool draws on results from computer models (called *predictive models*) that estimate the hydrodynamic and ecological effects that risk-reduction projects can have on asset damage and the effects of restoration projects on land building. Effects were considered for a range of risk-reduction, landscape, and ecosystem-service metrics and were made for two different environmental scenarios: moderate and less optimistic. The less optimistic scenario assumed higher sea-level rise and subsidence rates along with more-frequent and more-intense hurricanes than for the moderate scenario.

Specifically, the predictive models estimated the effects of risk-reduction projects on residual damage at three recurrence intervals (50, 100, and 500 years) across 56 communities in coastal Louisiana. Similarly, the models estimated the effects of restoration projects on 14 ecosystem-service metrics across 12 regions in coastal Louisiana. The Planning Tool also evaluated the effects of projects and alternatives on 11 additional decision criteria, such as *support of navigation* and *use of natural processes*, using project-specific information along with the risk-reduction and ecosystem-service effects of the projects.

Figure S.1
Locations of Restoration Projects Evaluated by the Planning Tool

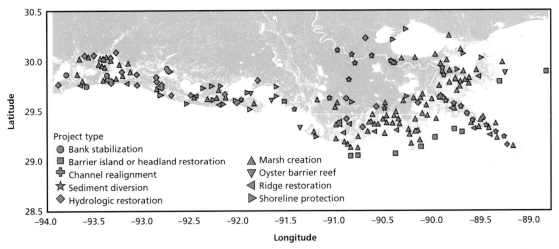

NOTE: Each symbol represents an individual project that may cover a much larger area than the symbol itself does, such as an entire parish.
RAND *TR1266–S.1*

Formulating Alternative Comprehensive Solutions

The Planning Tool identifies *alternatives* (groups of projects) over a 50-year planning horizon using an optimization model. The Planning Tool uses a *mixed-integer program* (MIP) to identify alternatives that minimize coast-wide risk to economic assets through risk-reduction projects and maximize coast-wide land building through restoration projects while satisfying a set of constraints. Specifically, an alternative's estimated costs cannot exceed available funding, sediment requirements cannot exceed available sediment resources, and river flow from diversions cannot reduce downstream flows below an acceptable level.

CPRA used the Planning Tool to iteratively develop and evaluate a large set of alternatives. For each iteration, the RAND team used the Planning Tool to formulate different alternatives. These results were provided to CPRA through an interactive, computer-based interface. CPRA then reviewed the analysis, shared selected results with its stakeholders, and provided the RAND team with revised specifications for additional alternatives.

This iterative process helped inform CPRA decisions about allocating funding between risk-reduction and restoration projects and the relative emphasis to place on near-term versus long-term benefits. Figure S.2, for example, shows estimates of long-term coast-wide land

Figure S.2
Long-Term Risk Reduction and Long-Term Land Building for Different Funding Splits and Funding Scenarios

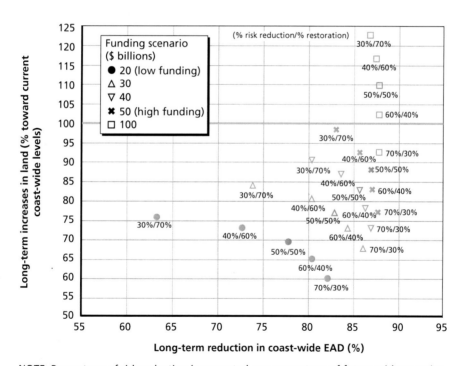

NOTE: Percentage of risk reduction is presented as a percentage of future without action (FWOA) expected annual damage (EAD) from flooding. EAD represents the monetary damage that would occur, on average, as a result of flooding from category 3 or greater storms in any given year, if a particular region were subjected to the same specific conditions and probability distribution of flood depths over many years. Land building is presented as a percentage of land lost under FWOA conditions. Long-term results are those for year 50. Symbols indicate different funding scenarios. Labels indicate different funding splits (risk reduction/restoration). Results are for the moderate scenario. Results for a 50/50 split are colored red.

RAND *TR1266-S.2*

area (vertical axis) and long-term coast-wide risk reduction (horizontal axis) for alternatives that differ in terms of total available funding (symbol) and different allocations between risk-reduction and restoration projects (labels and coloring). This figure helped CPRA decide to develop the Master Plan around a $50 billion budget and to allocate funding equally to risk-reduction and restoration projects.

Deliberating over Alternatives to Develop the Master Plan

RAND developed several versions of a visualizer of Planning Tool results to support the Master Plan deliberations. Each version contained specific visualizations based on a set of Planning Tool evaluations stored in an internal database. These visualizations were used to support numerous workshops with stakeholders and meetings with CPRA management and other key decisionmakers.

CPRA used the Planning Tool to support its selection of the specific alternative that serves as the foundation of the 50-year, $50 billion 2012 *Louisiana's Comprehensive Master Plan for a Sustainable Coast*. The draft Master Plan (CPRA, 2012a) was released in January 2012 for public review and comment. CPRA subsequently held three all-day public meetings and more than 50 meetings with community groups, parish officials, legislators, and stakeholder groups. CPRA then used the Planning Tool to reformulate alternatives based on revised project information and input from public comments. This information helped develop the final Master Plan (CPRA, 2012c), which was presented to the Louisiana legislature in April 2012 and passed into law in May 2012.

The 2012 Master Plan

The 2012 Master Plan is the first comprehensive solution for Louisiana's coast to receive broad support from the Louisiana public and the many agencies, federal, state, and local, engaged in protecting the Gulf Coast. It is based on $50 billion of funding (in 2010 dollars) over the next 50 years allocated broadly across the coast and among different project types (Figure S.3).

The Planning Tool estimates that implementation of the Master Plan would dramatically decrease coast-wide flood risk from a currently estimated level of $2.4 billion on average today to between $2.4 billion and $5.5 billion in year 50 with the full implementation of the Master Plan (Figure S.4). Without the Master Plan in place, EAD could exceed $23 billion under the less optimistic scenario.

The Planning Tool also estimates that the Master Plan, under moderate assumptions, would stabilize the coastal land area by around 2040 and increase land thereafter (Figure S.5). Under less optimistic assumptions, however, coast-wide land area never stabilizes, and land loss would be severe (Figure S.6). This result suggests that it will be critical to adapt the Master Plan if sea level rises and other key conditions are less favorable than those in the moderate scenario.

The Planning Tool played a critical role in the development of CPRA's Master Plan by providing information to support the deliberation needed to formulate a single 50-year plan. It provided a structured, analytic framework for comparing different risk-reduction and restoration projects, formulating many different alternatives, each representing one possible comprehensive approach to solving the coast's flood risk and land-loss problems. The resulting 50-year Master Plan received strong public support and passed the Louisiana legislature unanimously in May 2012.

Figure S.3
Master Plan Funding, by Project Type (millions of 2010 dollars)

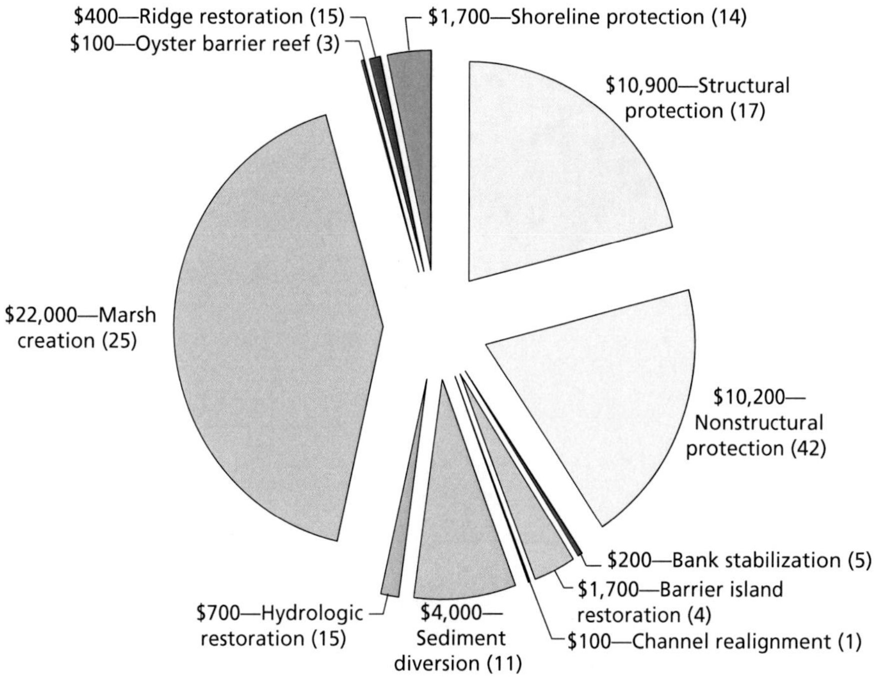

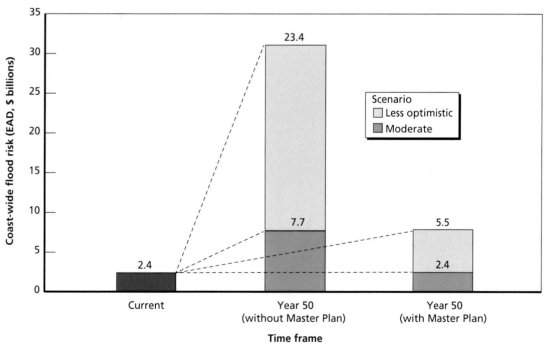

$400—Ridge restoration (15)
$100—Oyster barrier reef (3)
$1,700—Shoreline protection (14)
$10,900—Structural protection (17)
$22,000—Marsh creation (25)
$10,200—Nonstructural protection (42)
$200—Bank stabilization (5)
$1,700—Barrier island restoration (4)
$100—Channel realignment (1)
$700—Hydrologic restoration (15)
$4,000—Sediment diversion (11)

NOTE: The numbers in parentheses indicate the number of projects of each type included in the Master Plan. Funding is rounded to the nearest $100 million

RAND *TR1266-S.3*

Figure S.4
Coast-Wide Flood Risk for Current Conditions, Year 50 Without the Master Plan, and Year 50 with the Master Plan for the Moderate and Less Optimistic Scenarios

Coast-wide flood risk (EAD, $ billions)
35
30
25
20
15
10
5
0
23.4
7.7
5.5
2.4
2.4
Scenario
Less optimistic
Moderate
Current
Year 50 (without Master Plan)
Year 50 (with Master Plan)
Time frame

RAND *TR1266-S.4*

Figure S.5
Change in Land Area With and Without the Master Plan for the Moderate Scenario

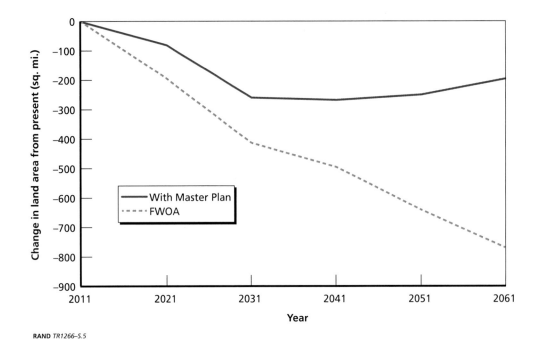

Figure S.6
Change in Land Area With and Without the Master Plan for the Less Optimistic Scenario

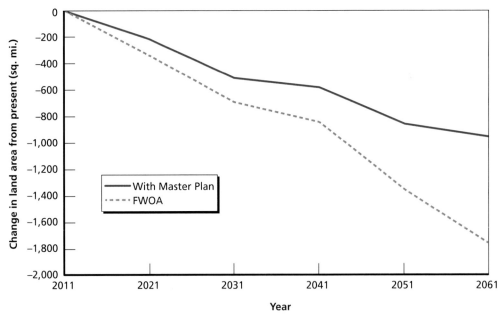

Acknowledgments

We would like to thank the staff of the Coastal Protection and Restoration Authority (CPRA) for their support throughout the Master Plan effort. We would especially like to thank Kirk Rhinehart, Natalie Snider, Karim Belhadjali, and Melanie Saucier of CPRA for their support and guidance. Members of the Planning Tool Technical Advisory Committee—John Boland, Benjamin F. Hobbs, and Leonard Shabman—the Master Plan's Science Engineering Board, and internal RAND reviewers have provided thoughtful reviews and helpful advice at various stages of development. Collaboration by our partners, Brown and Caldwell and the University of New Orleans, has been greatly appreciated; Cindy Paulson, Joanne Chamberlain, Alaina Owens, Joe Wyble, and Stephanie Hanses of Brown and Caldwell and Denise J. Reed of the University of New Orleans have been especially helpful throughout the process. We have worked closely with Jordan Fischbach and David R. Johnson at RAND to ensure that the results from the flood risk modeling were appropriately used in the Planning Tool. Finally, we would like to thank Anna Smith of RAND and Keith Crane, director of RAND's Environment, Energy, and Economic Development Program for their assistance throughout the effort.

Abbreviations

cfs	cubic feet per second
CPRA	Coastal Protection and Restoration Authority
csv	comma-separated values
EAD	expected annual damage
EEED	Environment, Energy, and Economic Development Program
FEMA	Federal Emergency Management Agency
FWOA	future without action
FWP	future with project
GAMS	General Algebraic Modeling System
GIWW	Gulf Intracoastal Waterway
Hazus	Hazards—United States
ISE	RAND Infrastructure, Safety, and Environment
LACPR	Louisiana Coastal Protection and Restoration
MCDA	multicriterion decision analysis
MIP	mixed-integer program
O&M	operations and maintenance
Planning Tool—TAC	Technical Advisory Committee
USACE	U.S. Army Corps of Engineers

CHAPTER ONE

Introduction

Coastal Louisiana is on an unsustainable trajectory of ongoing conversion of coastal land to open water and increasing hurricane flood risk. The causes of the ongoing land loss are varied and include natural and human-caused land subsidence, rising sea level, and the loss of nourishing sediment from Mississippi River flows that is now deposited deep in the Gulf of Mexico. Since the 1930s, 1,800 square miles have been lost (Couvillion et al., 2011). Without major investments in coastal restoration, the Coastal Protection and Restoration Authority of Louisiana (CPRA) estimates that an additional 800 square miles of coastal land could be lost over the next 50 years under moderate assumptions about future conditions, and 1,800 square miles under less optimistic assumptions (CPRA, 2012a).

This loss of land is changing the nature of the coastal environment profoundly and diminishing many of its benefits, including habitats for commercially and recreationally important species. Land loss is also increasing hurricane flood risk because coastal land provides the first line of defense against storm surge. As tragically demonstrated by the flooding and levee failures caused by Hurricane Katrina and later damage from Hurricane Rita in 2005, many of Louisiana's residents and commercial and business establishments face high levels of risk to hurricane storm-surge flooding. Hurricane Katrina, for example, inflicted $8 billion to $10 billion in direct damage to New Orleans residences alone, with 200,000 homes and 15,000 apartment units destroyed in the city (Grossi and Muir-Wood, 2006; Brinkley, 2006).

CPRA estimates that Louisiana currently faces an average of $2.4 billion of damage annually just to residences, commercial buildings, and industrial structures.[1] As communities and economic assets grow during the coming decades, the land that provides a protective buffer is anticipated to continue to degrade. Sea-level rise and subsidence rates may accelerate (Vermeer and Rahmstorf, 2009; Kolker, Allison, and Hameed, 2011), and hurricanes may increase in frequency and magnitude in response to a changing climate (Knutson et al., 2010). As a consequence, annual damage is expected to rise without investment in risk-reduction and restoration projects. Under moderate estimates of future demographic and economic changes, sea-level rise, subsidence, and changes in hurricanes, expected damage could increase to $7.7 billion per year in 50 years. Under less optimistic estimates of future conditions, EAD could exceed $23 billion in 50 years.

To address this challenge, CPRA developed *Louisiana's Comprehensive Master Plan for a Sustainable Coast* (CPRA, 2012c) to define a 50-year plan for reducing hurricane flood risk and

---

[1] Expected annual damage (EAD) represents the monetary damage that would occur on average as a result of flooding from category 3 or greater storms in any given year, if a particular region were subjected to the same specific conditions and probability distribution of flood depths over many years. In a given year, such as one in which a large hurricane makes landfall, damage amounts would be much larger than the average EAD. In other years, no damage would occur.

achieving a sustainable landscape. As part of this effort, CPRA supported the development of a computer-based decision-support tool called the Planning Tool to (1) make analytical and objective comparisons of hundreds of different risk-reduction and restoration projects, (2) identify and assess groups of projects (called *alternatives*) that could make up comprehensive solutions, and (3) display the trade-offs interactively to support iterative deliberation over alternatives. This document describes the Planning Tool and its use to support the development of the Master Plan.

Planning Objectives

The Master Plan defined five primary objectives and sought to develop a solution that meets each of them:

1. Reduce economic losses from storm surge–based flooding to residential, public, industrial, and commercial infrastructure.
2. Promote a sustainable coastal ecosystem by harnessing the natural processes of the system.
3. Provide habitats suitable to support an array of commercial and recreational activities coast-wide.
4. Sustain the unique heritage of coastal Louisiana by protecting historic properties and traditional living cultures and their ties and relationships to the natural environment.
5. Promote a viable working coast to support regionally and nationally important businesses and industries.

To meet each of these objectives, long-standing trends of increased flooding from hurricanes and accelerated losses of coastal land will need to be slowed and ultimately reversed. Increasing flood risks threaten the economic vitality and cultural heritage of coastal Louisiana. Loss of the coastal landscape threatens the viability of the wide array of ecosystem services that are currently provided by the Louisiana coastal region.

Thus, the focus of the Master Plan was to demonstrate a way *to reduce future expected annual hurricane-surge flood damage* and *stabilize coastal land area over the coming decades.* These two factors, called *decision drivers*, are the foundation for measuring success in the Louisiana coastal region. The Master Plan is intended to demonstrate how to achieve progress toward both of these goals in the long term (over the next 50 years), as well as the near term (over the next 20 years).

Planning Under Uncertainty

The Master Plan is designed to achieve coastal sustainability in the long-term future, even though the specific nature of the future is unknown. Scientists have developed a wide range of credible estimates of how factors affecting coastal conditions could change. CPRA strived to develop a Master Plan that is robust to as much uncertainty about these future conditions as possible. Robustness can be achieved in two steps: (1) by identifying near-term investments that will perform sufficiently well over a wide range of future conditions and (2) determining which other investments can be implemented successfully at later points in time, depending on how the future unfolds and in response to new or improved information. The Master Plan thus provides a set of near-term investments to make in the next 20 years. It also specifies additional investments to be made during the subsequent 30 years. The precise order of implementation within the two time periods and the specific projects in the later period will need to be adjusted over time. Such an adaptive Master Plan can best ensure that the state achieves its goals despite the uncertainties of the future.

Purpose of the Planning Tool

The Planning Tool was developed over several years by a team of researchers at the RAND Corporation, guided by CPRA's Master Plan Delivery Team.[2] Its development was overseen and reviewed by a Technical Advisory Committee (Planning Tool—TAC) made up of three experts in coastal and natural resource planning.[3]

The Planning Tool helped CPRA to develop a consistent, scientific base of information to support three sets of deliberations leading to the final Master Plan:

1. *Comparison of individual risk-reduction and restoration projects:* Which flood risk-reduction and restoration projects are most consistent with the objectives of the Master Plan?
2. *Formulation of alternatives made up of individual projects:* What groups of projects (or alternatives) can be implemented over a 50-year period to best achieve the objectives of the Master Plan given constraints on funding, sediment resources, and river flow?
3. *Comparison of alternatives based on the assumptions of additivity of projects' effects on coast-wide outcomes and independence between risk-reduction and restoration projects:* When compared across all the objectives of the Master Plan, which alternative is preferred?

A fourth analysis, evaluation and comparison of integrated alternatives, was completed after the publication of the Master Plan and is also described in this report.

In the following chapters, we describe the methodology and assumptions underlying the Planning Tool, its analytical procedures, and results for each step of the analysis.

---

[2] The Master Plan Delivery Team was made up of CPRA planners and selected members of the consulting team from RAND, Brown and Caldwell, and the University of New Orleans.

[3] The Planning Tool—TAC consisted of John Boland and Benjamin Hobbs of Johns Hopkins University and Leonard Shabman of Resources for the Future.

CHAPTER TWO

Model Description and Assumptions

The Planning Tool identifies *alternatives* (groups of projects) over a 50-year planning horizon using an optimization model. These alternatives (1) minimize coast-wide risk to economic assets through risk-reduction projects and (2) maximize coast-wide land building through restoration projects. Risk-reduction projects include structural features, such as levees and floodwalls, and nonstructural programs that reduce flood damage to residential and commercial structures through elevating, flood-proofing, or removing the structures. Restoration projects include bank stabilization, barrier island restoration, channel realignment, sediment diversion, hydrologic restoration, marsh creation, oyster barrier reef, ridge restoration, and shoreline protection. (See CPRA, 2012c, Appendix C, for details on all the projects considered.)

The mathematical statement that combines these decision drivers of risk reduction and coastal restoration is called an *objective function*. Each alternative also satisfies a series of *constraints*. These constraints take several forms. Some constraints ensure that the costs of constructing, operating, and maintaining the alternative do not exceed expected funding available for risk-reduction and restoration projects. Others ensure that available sediment for mechanical land building is not exceeded and that the diversion flow capacity of rivers for diversions and channel realignments is sufficient. Some constraints prevent inclusion of multiple projects that may be mutually exclusive. Other constraints reflect state and stakeholder preferences for achieving the Master Plan goals in other forms.

Predictive Modeling Framework

The Planning Tool was designed to support the Master Plan process by formulating many different alternatives, drawing on results from computer models that estimate the hydrodynamic and ecological effects of risk-reduction projects on asset damage and the effects of restoration projects on land building or loss (Figure 2.1) (see CPRA, 2012c, Appendix D). These are also known as *process effect models* and, in the Master Plan, *predictive models*. For consistency, we use the term *predictive models* in this document. In a process separate from the development of the Planning Tool, these predictive models were developed to estimate the effects that each individual project would have over 50 years relative to conditions in a *future without action* (FWOA).[1] Effects were considered for a range of risk-reduction, landscape, and ecosystem-service metrics and were made for two different environmental scenarios—moderate and less optimistic—discussed later in this chapter.

---

[1] See CPRA (2012c, Appendix D) for more detail about the specific linkages and interactions among the models.

Figure 2.1
Linkages and Feedbacks Among Predictive Models

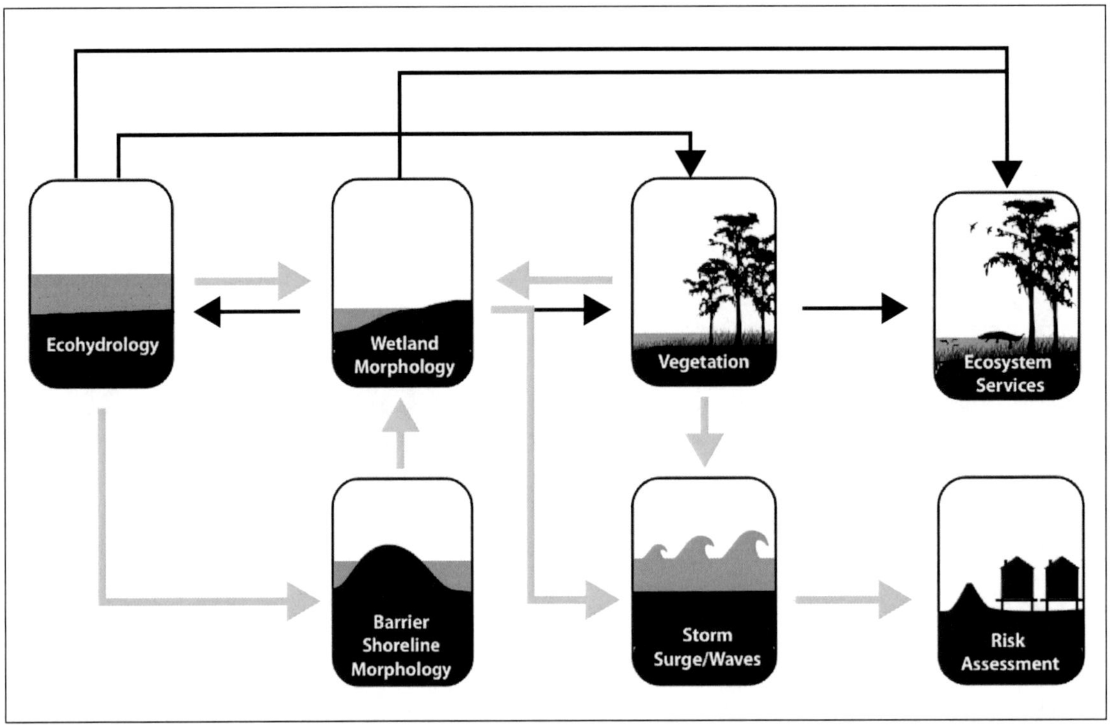

SOURCE: CPRA, 2012c, p. D-5.
NOTE: Linkages new to the Master Plan are indicated in orange.
RAND *TR1266–2.1*

Formulation of Alternatives

Each alternative identified by the Planning Tool can be thought of as the answer to a specific question, such as one of the following:

- What set of projects would build the most land and reduce the most risk coast-wide by 2050 with $25 billion available for risk reduction and $25 billion available for restoration projects?
- How would the alternative developed above differ if the state favored projects making the most use of natural processes or providing the greatest benefit to navigation?
- What would be the impact of such an alternative on the wide range of ecosystem-related metrics and levels of risk faced by communities across the coast?
- How would the choice of projects differ if sea-level rise and other factors were more extreme than those in the moderate scenario?
- How would the choice of projects differ if the relative emphases on near-term and long-term goals were shifted?

Basis of the Approach in Decision Theory

The decision analytic approach supported by the Planning Tool is grounded in decision theory. At its core, the Planning Tool is designed to support a *deliberation-with-analysis process* by which quantitative analysis is used not to provide a single answer but rather to frame and illuminate key policy trade-offs (National Research Council, 2009).

The Planning Tool supports such a process by producing information about project selection and potential effects under an assumed set of inputs reflecting different preferences and scenarios reflecting expectations about the future. Such an exploratory modeling approach is suited for long-term policy questions in which uncertainty is significant, there are a variety of views on desirable outcomes, and there is disagreement about how the system will respond to future stressors (Lempert, Popper, and Bankes, 2003).

The Planning Tool seeks to define alternatives that maximize the goals of the Master Plan while satisfying a wide range of constraints. Earlier versions of the Planning Tool relied heavily on multicriterion decision analysis (MCDA) (Keeney and Raiffa, 1993; Lahdelma, Salminen, and Hokkanen, 2000; Kiker et al., 2005; Linkov et al., 2006) as a structured approach to defining alternatives that conformed to a set of preferences, as reflected by a corresponding set of weights. Specifically, in its earlier form, the Planning Tool's mixed-integer program (MIP) employed a weight-based application of multiobjective programming to deal with its multiple, competing objectives and a constrained decision space.[2] Although theoretically attractive, such an approach was deemed to not be implementable for several reasons:

- The metrics that would form the basis of decision criteria were not easily placed on a consistent scale for comparison.
- The number of potential criteria (including more than ten ecosystem-service metrics) was large, and combining them in a single-value function was viewed as too complex to sufficiently communicate to stakeholders.
- The interpretation of weights for each factor in the objective function did not have a straightforward interpretation for CPRA or its stakeholders.

The current version of the Planning Tool continues to use a standard mixed-integer programming approach (Schrijver, 1998) but with a simplified application of MCDA to solve the constrained optimization problem of maximizing a simple multicriterion objective function subject to funding and other constraints. The current approach continues to use elements of multiobjective programming but with a focus on the constraint-based approach to dealing with multiple objectives (Romero, 1991). Rather than including all decision criteria within the MIP's objective function as originally envisioned, the Planning Tool uses a simple and easily understood objective function made up of only near-term and long-term risk reduction and land building. From here forward, risk reduction and land building are therefore referred to as decision drivers. All other decision criteria are used by the MIP as constraints. Alternatives are selected on the basis of whether they perform sufficiently well across a broad range of outcomes.

---

[2] Multiobjective programming is an approach to MCDA that generates solutions that are members of the set of Pareto-efficient solutions for an optimization problem defined by multiple objectives subject to a constrained decision space (Romero, 1991).

Due to time limitations imposed by the legislative calendar, not all capabilities of the Planning Tool were fully used to support the development of the Master Plan. For example, as described in "Predictive Modeling Framework" earlier in this chapter, all analyses used to formulate alternatives were based on the assumptions that project effects are additive and independent between risk-reduction and restoration projects. Also, alternatives were formulated on the basis of only two scenarios describing uncertain future conditions. The performance of the Master Plan could be significantly different from what one might expect if future conditions do not resemble one of the two scenarios. The Planning Tool should be used to more thoroughly test the robustness of the Master Plan under other scenario conditions and make adjustments accordingly.

Objective Function and Developing Alternatives Using Optimization

The Planning Tool uses an MIP to solve a *constrained optimization* problem identifying an alternative (i.e., group of projects) that provides the highest value of the objective function while satisfying all the constraints.[3] The Planning Tool's objective function has four basic terms: two decision drivers—risk reduction and land building—each at two points in time— 20 years and 50 years from the initiation of the Master Plan. These decision drivers reflect the Master Plan's overarching objectives as affirmed by stakeholders and local leaders.

Risk-Reduction Decision Driver

The Planning Tool takes into account the uncertainty of when and where floods will occur. Communities may go years without a serious flood, they may experience minor floods, or they may be severely flooded several years in a row—any number of variations is possible. Risk reduction is thus defined in terms of reduction in EAD—that is, the average damage that would be expected due to hurricane storm-surge flooding and waves in a particular year (e.g., year 50) across a statistical range of possible flooding events that could happen in that year. These averages are expressed as dollars in damage per year and do not imply that every community will flood every year. Note that *flood risk* in this context refers only to the direct economic flood damage to structures and does not include loss of life or indirect economic impacts of flooding.

Reductions in EAD are calculated relative to risk under the future without action. In the future without action, CPRA assumes that no new projects will be undertaken beyond those already authorized and funded in 2012. The algorithm used to calculate each project's (or alternative's) risk-reduction score is based on the percentage of total EAD under FWOA conditions that is eliminated for each community when a project or alternative is implemented. A coast-wide level of risk reduction is calculated using a weighted average across communities of the percentage of total EAD under a future without action that is eliminated. The weighted average ensures that each dollar of EAD reduction is equally valuable across all communities. Reductions in EAD are assumed to be additive across projects and are capped at complete elimination of risk for each community.

---

[3] A MIP is required because the optimization model must be able to find solutions using binary (0 or 1) decision variables that represent whether a project is in or out of the solution and using continuous variables, such as the availability of funds or sediment. These constraints are discussed later in this chapter.

Land-Building Decision Driver

The second decision driver, land building, reflects the general positive relationship between both the amount of coastal land and flood risk reduction and the amount of coastal land and provision of ecosystem services in coastal Louisiana. It is measured simply in terms of the change in total land area coast-wide due to the implementation of restoration projects. This decision driver is calculated at the coast-wide level, and it is assumed that land is equally valuable across the coast. The Planning Tool assumes that the land-building effects of individual projects are additive. This approach allows the building of land in one region of the coast to compensate for loss of land in another region of the coast.

Objective Function

A simplified form of the objective function is shown in Expression 2.1.[4]

Let d_j represent the weight for decision criterion j, such that

$$Max \begin{pmatrix} d_1 \left(\text{alternative near-term reduction in EAD} \right) \\ +d_2 \left(\text{alternative long-term reduction in EAD} \right) \\ +d_3 \left(\text{alternative near-term coast-wide land area} \right) \\ +d_4 \left(\text{alternative long-term coast-wide land area} \right) \end{pmatrix},$$

2.1

where *near-term* refers to outcomes in year 20 and *long-term* refers to outcomes in year 50. Risk-reduction benefits are expressed in the form of reduction in EAD, and land-building benefits are expressed in the form of square miles of land. The weighting terms d_1, d_2, d_3, and d_4 are included to enable decisionmakers and stakeholders to specify the relative value they place on these four terms in Expression 2.1; the weights must sum to 1.[5] Exploring the influence of these relative weights is discussed in Chapter Four. Each of the four decision-driver scores for an alternative included in the objective function in Expression 2.1 is the sum of the corresponding decision-driver scores for the projects comprising the alternative, as shown in Equations 2.2 through 2.5.[6]

Decision variables indicate whether a particular project is started during a particular implementation period for a given alternative. A project is not included in an alternative if it is not started during any of the implementation periods under consideration. The decision variables, denoted by the symbol x, have values of either 0 (meaning the project is not started in the

---

[4] The modified objective function shown is included only to provide the reader with the general idea of the objective function. In the formal mathematical expression of the objective function, land-area benefits are expressed as a ratio that represents progress toward building the amount of land lost between current conditions and FWOA conditions from restoration projects only. Similarly, risk-reduction benefits are expressed as a ratio that represents progress toward eliminating FWOA EAD from risk-reduction projects only.

[5] The optimization problem is structured so that the decision variables related to reduction in EAD are independent of the decision variables related to land building. As such, the value of weights d_1 and d_2 do not affect the selection of restoration projects, and the value of the weights d_3 and d_4 do not affect the selection of risk-reduction projects. The value of the weight d_1 relative to the value of weight d_2 does, however, affect the solution, as does the value of the weight d_3 relative to the value of weight d_4. The relative value of these two groupings of weights does not affect which projects are selected for inclusion in an alternative.

[6] A set of linear constraints is applied to an alternative's long-term reduction of residual damage to cap the total progress in a single community at 100 percent because residual damage cannot fall below 0.

given implementation period) or 1 (meaning the project is started in the given implementation period). In mathematical terms, the decision variables are defined for each project type and implementation period. The symbol p_r is used to represent a member of the set of risk-reduction projects, p_e represents a member of the set of restoration projects, and i represents a member of the set of possible implementation periods. A decision variable value of 1 implies that the given project is started in implementation period i. Thus,

$$\text{alternative near-term reduction in EAD}$$
$$= \sum_{p_r} \sum_i \left(\text{near-term reduction in EAD}_{p_r,i} \times x_{p_r,i} \right),$$
\hfill 2.2

$$\text{alternative long-term reduction in EAD}$$
$$= \sum_{p_r} \sum_i \left(\text{long-term reduction in EAD}_{p_r,i} \times x_{p_r,i} \right),$$
\hfill 2.3

$$\text{alternative near-term coast-wide land area}$$
$$= \sum_{p_e} \sum_i \left(\text{near-term coast-wide land area}_{p_e,i} \times x_{p_e,i} \right),$$
\hfill 2.4

and

$$\text{alternative long-term coast-wide land area}$$
$$= \sum_{p_e} \sum_i \left(\text{long-term coast-wide land area}_{p_e,i} \times x_{p_e,i} \right).$$
\hfill 2.5

The symbol Σ denotes the summation of the individual terms to its right identified by their subscripts.

The Planning Tool adjusts project effects and costs to account for the time period in which projects are implemented. If a project is selected for implementation in the second period, for example, then its costs and effects will not have any bearing on the first period. Costs and effects are both shifted to begin later in the 50-year planning time horizon to correspond with the project being selected for implementation in the second period.

The Planning Tool calculates near-term (year 20) risk-reduction benefits using assumptions specific to the type of project (structural or nonstructural) and when construction of the project is completed. If a structural risk-reduction project is fully constructed by year 20, then the full risk-reduction benefits (as estimated at year 50) of the project are assumed to be realized in the near term. If the project is not fully constructed by year 20, then benefits of the project are 0 in the near term. Different assumptions are made for nonstructural projects. In the near term, benefits are assumed to accrue linearly between the year in which a project starts and the year in which the project is completely implemented. Projects that are completed by year 20 are assumed to provide the full benefits in year 20. Projects that are only partially completed by year 20 are assumed to provide a fraction of the full benefits equal to the percentage of the project constructed by year 20.

Through the optimization process, the Planning Tool identifies different alternatives consistent with the Master Plan objectives and specifies the time periods in which risk-reduction projects and restoration projects would be implemented.[7] Table 2.1 shows the breakdown of the three time periods the Planning Tool considers when selecting projects for implementation.

Metrics and Decision Criteria

The Planning Tool considered how projects and alternatives would affect a set of risk-reduction and ecosystem-service metrics. Specifically, the predictive models estimated the effects that risk-reduction projects would have on residual damage at three recurrence intervals (50, 100, and 500 years) across 56 communities in coastal Louisiana. The predictive models also estimated the effects that restoration projects would have on 14 ecosystem-service metrics across 12 regions in coastal Louisiana.

The Planning Tool also evaluated the effects of projects and alternatives on 11 additional decision criteria, such as support for navigation and use of natural processes, using project-specific information along with the risk-reduction and ecosystem-service effects of the projects.

The Planning Tool uses these metrics and decision criteria in two ways:

- *Project comparison and alternative formulation:* Metrics and decision criteria that could be calculated for individual projects were used to compare projects and formulate alternatives.
- *Detailed reporting of alternatives:* Some decision criteria could be scored only for an alternative and therefore were developed only for final reporting.

Metrics

Master Plan objective 1 (see p. 2) is represented in the Planning Tool in the form of three risk-reduction metrics, in addition to EAD. Each metric represents the reduction in residual damage for a specific storm-surge flood recurrence interval (50-, 100-, or 500-year recurrence),[8] all in 2010 constant price dollars:

- reduction in residual damage at the 50-year storm-surge flood recurrence interval
- reduction in residual damage at the 100-year storm-surge flood recurrence interval
- reduction in residual damage at the 500-year storm-surge flood recurrence interval.

Each metric is used to measure reduction in residual damage due to a project or alternative for communities specified to have a target level of protection for the respective storm-surge

---

[7] Note that the objective function of the Planning Tool is not spatially explicit and reduces to a single value representing coast-wide risk reduction and coast-wide increases in land.

[8] Each metric represents the difference in a recurrence interval's damage exceedance—the level of damage one would expect to surpass only with the probability associated with the given recurrence interval—for a future without action and the damage exceedance for with-project conditions. For example, the "reduction in residual damage at the 50-year recurrence interval" metric represents the difference between the level of damage under a future without action for which we would expect damage of that level or greater to occur with a probability of 2 percent and the level of damage under with-project conditions for which we would expect damage of that level or greater to occur with a probability of 2 percent.

Table 2.1
Time Periods Used for Allocating Funding over 50 Years and Calculating Near-Term and Long-Term Benefits

Time Period	Years	Target Years for Calculating Near- and Long-Term Benefits
1	2012 to 2031	Near term: year 20 (2031)
2	2032 to 2051	Long term: year 50 (2061)
3	2052 to 2061	

flood recurrence interval. Each of the 56 communities was targeted for 50-, 100-, or 500-year levels of protection.

In addition to the land-area decision driver, Master Plan objective 3 is represented in the Planning Tool in the form of 14 ecosystem-service metrics. Nine of these metrics were evaluated for each restoration project and were considered by the Planning Tool as alternatives were formulated:

1. alligator (habitat suitability units)[9]
2. oysters (habitat suitability units)
3. shrimp (habitat suitability units)
   a. brown shrimp (habitat suitability units)
   b. white shrimp (habitat suitability units)
4. saltwater fisheries (habitat suitability units)
5. waterfowl (habitat suitability units)
6. carbon sequestration (metric tons)
7. freshwater availability (suitability units)
8. nutrient uptake (kilograms)
9. storm surge and wave attenuation (suitability units).

Additional ecosystem-service metrics (crawfish, freshwater fisheries, other coastal wildlife, agriculture, and nature-based tourism) were not used by the Planning Tool to formulate alternatives but were displayed alongside the other metrics in the Planning Tool for comparison purposes only.

These metrics are described in the Master Plan (CPRA, 2012c, Appendix D).

Decision Criteria

Eleven additional decision criteria were defined to reflect other aspects of the Master Plan's five objectives. Each additional criterion relates to a specific Master Plan objective and was calculated or estimated for each relevant project using some combination of project attribute data, estimates from the predictive models, and expert judgment.

---

[9]   The predictive models calculate habitat suitability units for a specific ecosystem service across the coast by first calculating habitat suitability index (HSI) scores for each gridded area of potential area. The HSI scores are then multiplied by the amount of area for each grid and then summed across all grid points to yield a total amount of habitat suitability units. For example, a 1,000 sq. kilometer area with perfect habitat (HSI = 1.0) would translate to 1,000 habitat suitability units (1,000 × 1.0).

Table 2.2 provides a description for each additional decision criterion. Note that the two primary decision drivers (reduction in EAD and land building), the three risk-reduction metrics, and the 14 ecosystem-service metrics are not included in Table 2.2. As a result, Master Plan objective 3 is not shown in Table 2.2 because it is reflected only by land building and the 14 ecosystem-service metrics. Subsequent sections describe when and how the different decision criteria are used, and CPRA (2012c, Appendix B) provides additional information on their formulation.

Distribution of Flood Risk Reduction Across Socioeconomic Groups

The *distribution of flood risk reduction across socioeconomic groups* decision criterion calculates a project's impact on the amount of EAD in census tracts classified as impoverished by the U.S. Census Bureau in the 2005–2009 American Community Survey poverty data (U.S. Census Bureau, 2012). The difference in EAD under FWOA conditions and in EAD under future-with-project (FWP) conditions is calculated for each impoverished census tract. The sum of the reduction in EAD across impoverished census tracts represents a project's effect with respect to this decision criterion.

Use of Natural Processes

Two decision criteria were created to represent the use of natural processes (one for risk-reduction projects and one for restoration projects). The separation into two decision criteria supports the assumption of independence in the selection of risk-reduction and restoration projects. Project scores for these two decision criteria represent a project's tendency to support the use of natural river flows and flooding, referred to as natural processes. Scores ranging from −1 to 1 were estimated by CPRA with expert input from the Framework Development Team for each project.[10] Scores for risk-reduction projects were based on whether or not the project impeded existing natural processes or hydrologic connections with a structural barrier. Scores for restoration projects were based on whether or not a project increased natural hydrologic patterns of the estuary in areas where they are currently limited or obstructed.

Sustainability

This decision criterion seeks to reflect the sustainability of land built by restoration projects. Sustainability is approximated by a simple measure of persistence of land: the degree to which land that is built 40 years after construction is present ten years later (50 years after construction). Specifically, this decision criterion is equal to the changes in land between the 50th and 40th years after construction is completed. Scores greater than or equal to 0 indicate that land is persisting after 50 years of operation.

Operations and Maintenance

This decision criterion is calculated for restoration projects and is the negative ratio of a project's annual O&M costs to its total costs for a 50-year planning horizon. Scores that are closer to 0 are better than scores that are negative.

---

[10] The Master Plan Framework Development Team was made up of 33 representatives from business and industry; federal, state, and local governments; nongovernmental organizations; and coastal institutions and met monthly for several years in support of the Master Plan (see CPRA, 2012b).

Table 2.2
Decision Criteria Reflecting Master Plan Objectives

Master Plan Objective	Decision Criterion	Focus	Relevant Project Type
1	Distribution of flood risk reduction across socioeconomic groups	How flood risk reduction is distributed between impoverished and nonimpoverished communities	Risk reduction
2	Use of natural processes	Using natural processes to advance risk-reduction goals	Risk reduction, restoration
	Sustainability	Sustainability through year 50 of the landscape	Restoration
	O&M	Percentage of O&M costs relative to planning, design, and construction costs	Restoration
4	Support of cultural heritage	Future conditions that support people's ability to live in their coastal communities and use ecosystem services and natural resources for work or recreation	Alternatives made up of risk-reduction and restoration projects
	Flood protection of historic properties	Improving protection of properties and districts determined to be of historic value	Risk reduction
5	Support of navigation	Benefits and impediments to the navigation industry, including shallow- or deep-draft sectors that operate in federally authorized channels	Risk reduction, restoration
	Flood protection of strategic assets	Improving protection of strategic assets	Risk reduction
	Support of oil and gas	Benefits to the oil and gas industry and infrastructure, as well as key communities for the workforce	Alternatives made up of risk-reduction and restoration projects
Not applicable	Critical landforms	Building land associated with the 16 landscape features identified by USACE in the LACPR technical report (USACE, 2009)	Restoration

NOTE: O&M = operations and maintenance. LACPR = Louisiana Coastal Protection and Restoration.

Support of Cultural Heritage

This decision criterion cannot be calculated for individual projects and is therefore not used for comparing individual projects or in formulating alternatives. Rather, this decision criterion is calculated for full alternatives only after they have been formulated by the Planning Tool. This decision criterion allows CPRA to make comparisons between the FWOA condition and the various alternatives that were formulated. Scoring of alternatives is based on levels of risk reduction to communities and the provision of natural resources within a reasonable distance of the community.

Flood Protection of Historic Properties

CPRA used data provided from the Louisiana State Historic Preservation Office (SHPO), Department of Culture, Recreation and Tourism, Office of Cultural Development, Division of Archaeology to identify 5,472 properties and 32 districts as historic and seeks to protect them to the level of a 50-year flood event. This decision criterion represents the difference in conditions between the future without action and the future with project in the number of historic properties that flood due to a storm flood event at the 50-year recurrence interval. For this decision criterion, a property is considered to have flooded if the estimated flood depth for its census block is greater than 6 inches. Properties that would have flooded under FWOA conditions but that do not flood when a project is implemented are considered to be protected by the given project. A project's score is the ratio of the number of properties protected to the total number of historic properties under consideration. Protecting a greater number of properties from flooding earns a higher score.

Support of Navigation

This decision criterion was created to reflect support of navigation and was applied to both risk-reduction projects and restoration projects. Scores represent a project's tendency to maintain the navigability of federally authorized waterways. Scores ranging from –1 to 1 were estimated for each project by CPRA with expert input from the Framework Development Team and the Navigation Focus Group. Scores for this decision criterion are compared separately for risk-reduction and restoration projects. Scores for risk-reduction projects were based on the addition of structures to waterways that could cause increased travel times. Scores for restoration projects were based on the extent of open water adjacent to channels used by barge traffic, the potential for sediment accumulation in authorized channels, and the effects that diversions would have on lateral flows within a navigable channel. Separation into two decision criteria supports the assumption of independence in the selection of risk-reduction and restoration projects.

Unlike the other decision criteria, the scores for support of navigation could not be used in an additive manner for the formulation of alternatives because of the difficulty of reflecting the type and magnitude of impact on navigation. Instead, each project's score is compared with a set of absolute threshold values to determine whether the project performs well enough with respect to its respective support of the navigation decision criterion to be included in an alternative.

Flood Protection of Strategic Assets

CPRA used data compiled from the Louisiana Governor's Office of Homeland Security and Emergency Preparedness, the Louisiana Department of Economic Development, the Louisiana Department of Environmental Quality, the Federal Emergency Management Agency (FEMA) Hazards—United States (Hazus) database, and the U.S. Energy Information Administration to identify 179 strategic assets (e.g., critical chemical plants, natural gas facilities, strategic petroleum reserves, power plants, petroleum refineries, ports and terminal districts, airports, military installations, other federal facilities).

This criterion is included to ascertain whether strategic assets are protected from a 50-year flood event. The Planning Tool calculates the difference in the number of strategic assets that flood because of a storm flood event at the 50-year recurrence interval from the FWOA and FWP conditions. The decision criterion embeds the assumption that an asset is flooded if the estimated flood depth for a census block is greater than 6 inches. Assets that flood under FWOA conditions but do not flood when a project is implemented are considered to be protected by that project. A project's score is the ratio of the number of assets protected to the total number of strategic assets under consideration. Protecting a greater number of strategic assets from flooding generates a higher score.

Support of Oil and Gas

This decision criterion cannot be calculated for individual projects and is therefore not used for comparing individual projects or in formulating alternatives. Rather, this decision criterion is calculated for full alternatives only after they have been formulated by the Planning Tool. This decision criterion allows CPRA to make comparisons between the FWOA condition and the various alternatives that were formulated. Scores are based on whether a formulated alternative supports the persistence of land and has the ability to reduce flood risks to communities with strong ties to the oil and gas industry.

Critical Landforms

This decision criterion represents the proportion of the total possible land building related to critical landforms that is attributable to a project. A critical landform is one of 16 landscape features defined by U.S. Army Corps of Engineers' (USACE's) LACPR technical report (USACE, 2009). Total possible land building related to critical landforms is calculated as the sum of land building by projects associated with any critical landform. Land building is measured as the difference between land area when the project is implemented and land area under FWOA conditions at year 50. This decision criterion embeds the assumption that a project's construction is complete prior to the start of the 50-year planning horizon such that its effects on land building begin on day 1 of the planning horizon (i.e., measures the land building associated with 50 years of operation of a project).

Constraints

The Planning Tool ensures that each alternative formulated satisfies a set of constraints. Specifically, an alternative's estimated costs cannot exceed available funding, sediment requirements cannot exceed available sediment resources, and river flow from diversions cannot reduce downstream flows below 200,000 cubic feet per second (cfs) (the minimum flow volume assumed by CPRA to limit any detrimental effects on navigation or drinking-water supplies).

Four types of constraints are used to formulate alternatives:

- *Financial and natural resource constraints:* total funding, the funding split between risk-reduction and restoration projects, sediment availability, allowable sediment diversion capacity, and allowable number of diversions for specific reaches of the Mississippi River
- *Mutually exclusive project constraints:* restrictions on implementation of projects that are variations of the same concept at the same location or conflict in some other way
- *Project inclusion and exclusion constraints:* specification of the inclusion or exclusion of specific projects to reflect other CPRA planning considerations not evaluated by the predictive models or the Planning Tool
- *Outcome constraints:* requirements that alternatives perform sufficiently well relative to specific metrics and decision criteria.

Financial and natural resource constraints are generally beyond the influence of the state. Given a total amount of available funding, the Planning Tool allows the user to specify what percentage of funding is allocated to risk-reduction projects versus restoration projects. The Planning Tool also deals with uncertainty in total funding by allowing the user to select different budget levels for which he or she would like to formulate an alternative. Mutually exclusive project constraints reflect the specific nature of the projects included in the analysis; in some cases, a project can be modified to become non–mutually exclusive. Project inclusion and exclusion constraints and outcome constraints generally reflect CPRA preferences. Table 2.3 describes the constraints used to formulate alternatives and the possible values that each could take.

Financial and Natural Resource Constraints

Financial and natural resource constraints used include funding available for restoration projects, funding available for risk-reduction projects, sediment available for use in project construction, and an acceptable amount of river flow that can be diverted from the Mississippi and Atchafalaya Rivers.[11] Inequalities 2.6 through 2.10 are used as resource constraints in the Planning Tool.

Let p_e represent a member of the set of restoration projects and p_r represent a member of the set of risk-reduction projects. Let t represent a member of the set of time periods analyzed in the 50-year planning horizon. Let z represent a member of the set of rivers (i.e., Mississippi or Atchafalaya). Let k represent a member of the set of river reaches in the river system. Let i represent a member of the set of implementation periods. Let s represent a member of the set of possible sediment sources.

$$\sum_{p_e} \sum_i \left(\text{cost}_{p_e,i,t} \times x_{p_e,i} \right) \leq \text{restoration funding}_t,$$
for all values of t,

2.6

$$\sum_{p_r} \sum_i \left(\text{cost}_{p_r,i,t} \times x_{p_r,i} \right) \leq \text{risk-reduction funding}_t,$$
for all values of t,

2.7

$$\sum_{p_e} \sum_i \left(\text{sediment required}_{p_e,i,t,s} \times x_{p_e,i} \right) \leq \text{sediment available}_{t,s},$$
for all values of t and s,

2.8

$$\sum_{p_e} \sum_i \left(\text{river flow diverted}_{p_e,i,z} \times x_{p_e,i} \right) \leq \text{river flow}_z,$$
for all values of z,

2.9

---

[11] Sediment constraints are adjusted using continuous decision variables to allow for unused sediment in one time period to be carried over for use in later time periods in the 50-year planning horizon.

Table 2.3
Constraints Used to Formulate Alternatives

Constraint Type	Constraint	Values
Financial and natural resource constraints	Total funding by time period (2012–2031, 2032–2051, 2052–2061); split of funding available for restoration projects and risk-reduction projects	50-year funding: $20 billion, $50 billion; various funding splits between risk reduction and restoration projects
	Sediment availability constrained separately for 18 different sediment sources by time period (2012–2031, 2032–2051, 2052–2061)	Each source is assigned a total amount of sediment available for each of the three time periods[a]
	Allowable sediment diversion capacity (based on river flow rates)	Mississippi: 773,243 cfs Atchafalaya: 620,000 cfs
	Allowable number of diversions per defined reach of the river (there are 7 reaches)[b]	2 (unless otherwise required by the project inclusion or exclusion constraints)
Mutually exclusive project constraints	Specific rules that identify mutually exclusive projects	
Project inclusion or exclusion constraints	Specific rules that specify that a project is to be included or excluded from an alternative	
Outcome constraints	Minimum outcome for long-term reduction of residual damage at 50-, 100-, and 500-year intervals	Determined as part of the alternative-formulation process
	Minimum outcome for long-term coast-wide ecosystem-service metrics	
	Minimum level for an alternative's decision-criterion scores	

[a] Eighteen sources of sediment have been identified as locations from which sediment can be taken for use in mechanical land building (e.g., marsh creation and ridge restoration). Some sources are replenishing over time, and others do not replenish. Specific information on sediment sources and availability is provided in CPRA (2012c, Appendix B).

[b] Seven river reaches were defined by CPRA based on the relative distribution of diversion projects along the Mississippi River. Four of the reaches are located on the east side of the Mississippi River, and three of the reaches are located on the west side of the Mississippi River.

and

$$\sum_{p_e} \sum_i \left(\text{river-reach indicator}_{p_e,k} \times x_{p_e,i} \right) \leq \text{allowable number of diversions}_k,$$
for all values of k.

2.10

Mutually Exclusive Project and Project Inclusion or Exclusion Constraints

Mutually exclusive project constraints take the form shown in Inequality 2.11.

Let $x_{1,i}$ and $x_{2,i}$ represent decision variables for two projects under consideration that are mutually exclusive of one another.

$$\sum_i x_{1,i} + \sum_i x_{2,i} \leq 1.$$

2.11

For example, mutually exclusive constraints are used to ensure that only a single non-structural project is selected for any given parish.

Project inclusion and exclusion constraints specify projects to be included or excluded in an alternative. These constraints reflect other CPRA planning considerations not captured by the predictive models or the Planning Tool. CPRA specified when these constraints should and should not be applied.

Outcome Constraints

The Planning Tool can specify that alternatives achieve a minimum value for a specific risk-reduction or ecosystem-service metric, as shown in Inequality 2.12.

$$\sum_{p}\sum_{i}\left(\text{metric}_{p,i} \times x_{p,i}\right) \geq \text{performance threshold,} \qquad 2.12$$

where *performance threshold* refers to a minimum level of achievement of the particular metric, such as shrimp.

In some cases, CPRA has specified that alternatives achieve a minimum value for particular decision-criterion scores. The general format for such a constraint is shown in Inequality 2.13.

$$\sum_{p}\sum_{i}\left(\text{decision-criterion score}_{p,i} \times x_{p,i}\right) \geq \text{performance threshold,} \qquad 2.13$$

where *performance threshold* refers to a minimum level of achievement of the particular decision criterion.

For example, suppose the Planning Tool derives an alternative based on maximizing long-term land building alone and that the resultant score for the alternative's *use of natural processes* decision criterion is 20. A performance threshold of 30 could be set using Inequality 2.13. An alternative developed using this constraint would score a 30 or higher for the *use of natural processes* decision criterion but would likely achieve this score at a cost of reduced land building. Figure 2.2 illustrates this trade-off.

The first alternative, shown in blue, maximizes land area only. The second alternative, shown in green, adds a constraint on use of natural processes (vertical red dashed line) to increase performance relative to this decision criterion. The "maximum" vertical bars show the maximum achievable value for each of the two decision criteria. The figure shows that, without a constraint on use of natural processes (i.e., the *maximize land* alternative), the maximum score for land area is achieved but the *use of natural processes* score falls well below the maximum value. The addition of a constraint on the score for use of natural processes (i.e., the *use of natural processes* alternative) reduces the land area achieved but increases the score for *use of natural processes* to at least the level of the constraint (though not necessarily all the way to the maximum achievable score).

Modeling Projects Under Different Scenarios

As a key part of its analysis of robustness, CPRA developed and evaluated projects and alternatives under different environmental and funding scenarios. A concurrent uncertainty analysis, focused on uncertainty in the formulation of the predictive models, was also performed by

Figure 2.2
Illustration of Two Alternatives and Their Scores Relative to Land-Area Use of Natural Processes

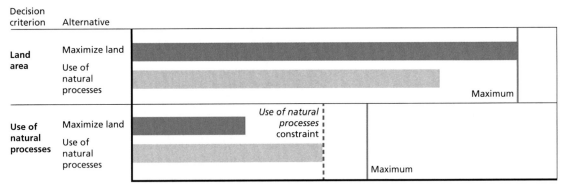

NOTE: The blue bars reflect results for the alternative that maximizes land. The green bars reflect results for the alternative that increases the *use of natural processes* decision-criterion score.
RAND TR1266-2.2

CPRA contractors but was not completed in time to be incorporated into the alternative-formulation process. As a result, the robustness analysis performed for the Master Plan relies on project effect estimates under the environmental and funding scenarios only.

Environmental Scenarios

The Master Plan developed two scenarios reflecting uncertainty in future physical conditions (see CPRA, 2012c, Appendix C, for details on the development and specifications of these scenarios) for use in formulating alternatives. The Planning Tool uses estimates of FWOA conditions and individual project effects on risk, landscape, and ecosystem-service metrics for each scenario.

Moderate Scenario

The predictive modeling teams used pertinent scientific literature and expert judgment to develop estimates of a moderate future condition for the following key factors:

- sea-level rise
- subsidence
- storm frequency
- storm intensity
- Mississippi River discharge
- rainfall
- evapotranspiration
- Mississippi River nutrient concentration
- marsh collapse threshold.

Less Optimistic Scenario

This scenario assumes less advantageous trends in the same factors described for the moderate scenario above.

Funding Scenarios

CPRA developed scenarios for funding per year over the next 50 years by evaluating a wide variety of funding sources and estimating a range for funding from each source over time. From these ranges, two scenarios were defined: a low-funding scenario (totaling about $20 billion over 50 years) and a high-funding scenario (totaling about $50 billion over 50 years). CPRA also developed additional scenarios, including a $100 billion funding scenario in which CPRA has $2 billion per year to spend on implementing the Master Plan. Table 2.4 shows the available funding, by time period, for the low-funding and high-funding scenarios.

Key Assumptions in the Development of Alternatives

This section summarizes the primary assumptions that underlie the calculations performed within the Planning Tool for the development of alternatives.

Risk-Reduction Projects Do Not Affect the Landscape or Ecosystem-Service Metrics, and Restoration Projects and Landscape Changes Do Not Affect Storm-Surge Risk

For purposes of the Master Plan, the Planning Tool includes the assumption that risk-reduction projects affect only flood risk and its related metrics and decision criteria. Similarly, restoration projects are assumed to affect only land area, ecosystem-service metrics, and related decision criteria. (See "Objective Function and Developing Alternatives Using Optimization" earlier in this chapter for discussion of metrics and decision criteria.) These assumptions apply when the Planning Tool is used to compare individual projects and when the Planning Tool is used to formulate alternatives, or groups of projects. These assumptions are necessary because of the current computational limitations of running the complex suite of predictive models, but they may bias the effects attributed to an alternative. Without accounting for the effects that land building can have on risk reduction, the estimates of risk reduction attributed to an alternative are likely to be underestimates. As a result, alternatives may be formulated that overprotect some areas of the coast. Similarly, land and ecosystem-service estimates may be biased upward without accounting for the effects that structural risk-reduction projects can have on the landscape and the ecosystem. For some aspects of the ecosystem, formulated alternatives may cause greater harm than what is estimated under this assumption.

The predictive models evaluated a small set of full alternatives in which all projects in a given alternative are modeled simultaneously to evaluate their combined effects on all metrics and decision criteria. This evaluation provides greater clarity regarding the implications of this assumption (see Chapter Four).

Physical and Biological Effects of Individual Projects Are Additive

The Planning Tool includes an assumption that the combined effects of two or more projects are additive. This assumption applies to flood risk metrics, land-area building, ecosystem-

Table 2.4
Funding Amounts ($ billions), by Time Period, for Two Funding Scenarios

Funding Scenario	2012–2031	2032–2051	2052–2061	Total
Low funding	8.9	7.3	3.9	20.1
High funding	26.0	15.3	9.0	50.3

service metrics, and most decision criteria. The Planning Tool team acknowledges that this assumption does not allow the Planning Tool to capture the synergy of projects or conflict between projects and thus could lead to biases when formulating alternatives. In some instances, this assumption may lead to an overestimate of the benefits attributed to an alternative. For example, the sum of benefits across multiple small diversions may be an overestimate of their combined benefit. Similarly, if two risk-reduction projects overlap in the areas they protect, this could also lead to an overestimate of the benefits attributed to an alternative. In other instances, this assumption may lead to an underestimate of the benefits attributed to an alternative. For example, the sum of the benefits of a sediment-diversion and a marsh-creation project that affect the same area of the coast may underestimate their combined effect because synergies may exist between the two projects.

Additional rules were developed to minimize the biases that could occur as a result of the assumption of additive project effects. Projects with too much overlap in their effects were designated as mutually exclusive if their combined benefit could not reasonably be considered additive. Also, limits were set on the number of diversions that could be specified to occur on a given reach of the Mississippi River.

The additive assumption also helps reduce the number of possible alternatives from a size that would be computationally infeasible to model. The assumption was removed when the projects within the Master Plan were modeled concurrently by the predictive models. The post–Master Plan estimates of future coast-wide flood risk and land area under the Master Plan by the predictive models were quite similar to the estimates from the Planning Tool based on the additive assumption. The additive assumption was more problematic when considering individual ecosystem-service metrics.

Funding Scenarios Are Known

Data provided to the Planning Tool team for estimating future funding over time are assumed to be reasonably reliable. Therefore, the Planning Tool uses these data to constrain the selection of projects that can be implemented in each period given the available funds. The Planning Tool is structured to allow for the same planning questions to be explored under multiple funding scenarios. This structure allows stakeholders to understand the impact that funding can have on planning decisions.

Funding Is Available for the Entire Implementation Period

The Planning Tool assumes that the entire amount of funding allocated for an implementation period is available at the start of the implementation period and remains available for use throughout that implementation period. This simplifying assumption was made under the guidance of CPRA and was determined to be appropriate given the potential uncertainty in funding flows.

Funding Cannot Be Saved for Use in Later Implementation Periods

The Planning Tool assumes that any funds unspent in an implementation period cannot be carried over for use in later implementation intervals. This assumption ensures that the Planning Tool's MIP develops appropriate results. By assuming that funds cannot be saved for use in later implementation periods, the MIP is prevented from saving funds to implement every project so that its optimal benefit occurs precisely at the two time intervals at which benefits are specifically measured (i.e., years 20 and 50). Assuming that funds cannot be saved also prevents the MIP from avoiding O&M costs by implementing projects as late as possible in the 50-year time horizon.

Projects Begin Planning and Design in the First Year of an Implementation Period

The Planning Tool divides the 50-year planning horizon into three implementation periods: years 1 through 20, years 21 through 40, and years 41 through 50. (To simplify the description of the selected projects, the Master Plan presents the second and third implementation periods as a single period.) The Planning Tool then evaluates in which, if any, of these three implementation periods a project should begin in order to allow the alternative to best achieve the Master Plan objectives. The Planning Tool assumes that a project begins its planning and design during the first year of the implementation period for which it was selected. This assumption reflects CPRA's desire to consider broad differences in implementation times while not overly constraining future sequencing of projects.

Project Effects Are Offset by Planning, Design, and Construction Time

To put all projects on a common timeline, the predictive models (see CPRA, 2012c, Appendix D) are structured such that the effects of a project begin immediately at the start of the Master Plan's implementation period (i.e., on day 1 in year 1). However, to account for differences in projects' planning, design, and construction times, the Planning Tool assumes that effects estimated by the predictive models can be offset by the number of years required to plan, design, and construct a project. For instance, if it were estimated that it would take seven years for a project to be planned, designed, and constructed, then the effect estimated for year 1 by the predictive models would be shifted by the Planning Tool to represent the project's effect in year 8. This assumption accounts for the delay in benefits that would exist because of planning, design, and construction (see CPRA, 2012c, Appendix A, for details on project timelines).

Projects Must Continually Operate

The Planning Tool assumes that, once a project begins its planning and design, it will be constructed and operated until the end of the 50-year planning horizon. For instance, a project cannot be planned and designed in years 1 through 5 and then start construction in year 20. Instead, it must begin construction (thus incurring construction costs) in year 6.

Handling and Processing of Data Within the Planning Tool

The Planning Tool incorporates multiple modeling environments and data-management software to allow it to perform the calculations described in the preceding section. All of these software platforms are commercially available. In this section, a short description of each software platform is provided.

MySQL Database

Multiple MySQL databases were built to house the data used by the Planning Tool in project comparison and alternative formulation. Each database stores the specific inputs required by the Planning Tool, as well as the corresponding results of project comparisons and alternative formulations. Inputs used by the Planning Tool are initially read into the MySQL database from comma-separated values (csv) files. Input data and results from the constructed data-bases can be accessed by the programs Analytica and Tableau (explained below) for use in data exploration and visualization.

Analytica Module

A 64-bit server license of Analytica extracts input data from the MySQL databases and pre-processes it to generate project-level scores for decision criteria and metrics. These scores are passed back to the MySQL database for data visualization in Tableau and for use in the linear MIP optimization model run in General Algebraic Modeling System (GAMS). Results from GAMS (which are also stored in the MySQL database) are extracted by Analytica and post-processed to calculate alternative-level scores for decision criteria and metrics. These scores are also passed to the MySQL database for data visualization in Tableau.

General Algebraic Modeling System Optimization Module

GAMS is a modeling environment that allows the user to create and solve mathematical prob-lems using optimization techniques. GAMS reads "include files" of data stored in the MySQL server after being preprocessed by Analytica.[12] These data contain all the necessary inputs for the linear MIP. GAMS utilizes a commercial solver (CPLEX) to perform a branch-and-bound search algorithm to solve multiple configurations of the planning problem.[13] A solution to an instance of the planning problem is provided as a set of binary indicators for each project iden-tifying for which implementation period, if any, a project is selected.[14]

Tableau Results Visualizer

Tableau is a business analytic software platform that supports interactive data visualization and that can be adjusted in real time to demonstrate changes in results given differences in assump-tions or input data. Tableau extracts both project-level and alternative-level scores for decision criteria and metrics from the MySQL database. The extracted data are then used to develop visualizations of trade-offs that inform the decisionmaking process.

The RAND team developed multiple versions of a Planning Tool results visualizer to support the Master Plan deliberations. Each version contained specific visualizations based on a set of evaluations of the Planning Tool, stored in an internal database. Figure 2.3 shows two screen captures of the public version of the tool.

---

[12] An include file is a text file separate from the main code being used that is prepared in advance and can be called by programming language when needed.

[13] Branch-and-bound search algorithms are commonly used to search for optimal solutions to discrete optimization prob-lems. Branch-and-bound algorithms search the solution space in an organized fashion using estimated performance thresh-olds to remove groups of suboptimal solutions.

[14] The full solution also includes the set of continuous variables that describe the sediment budget over time for each sedi-ment source, but the primary point of interest is the binary indicators identifying for which implementation period, if any, a project is selected.

Figure 2.3
Two Screen Shots of the Public Version of the Planning Tool Results Visualizer

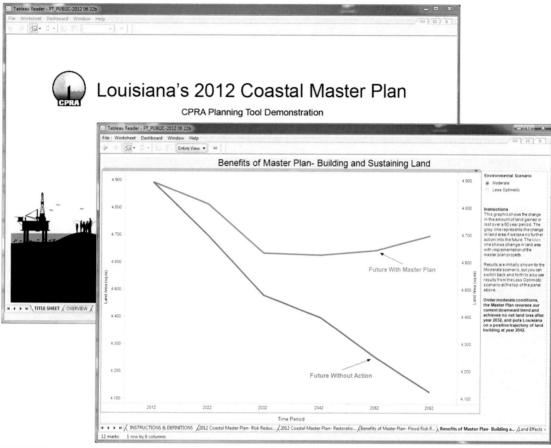

CHAPTER THREE

Analytic Procedures

This chapter describes the five analytic procedures required to develop and compare the alternatives identified in the Master Plan. CPRA, its consulting team, and the predictive modeling teams executed the first two procedures—characterization of projects and modeling project effects. Outputs from these procedures were then provided to the Planning Tool team along with input data on restoration and risk-reduction projects. In the third and fourth procedures—comparison of individual projects and formulation of alternatives—the Planning Tool was used to inform the development of the Master Plan. The last—integrated evaluation of alternatives—is made up of additional analyses of a few alternatives by the predictive models (as distinguished from the first-order modeling of individual project effects) and reevaluation of the results of the modeling of the alternatives using the Planning Tool.

Characterization of Projects

As described in CPRA (2012c, Appendix A), the Master Plan evaluated structural risk-reduction and restoration project concepts that were developed prior to the start of the Master Plan process. The Master Plan Delivery Team also developed a comprehensive suite of nonstructural projects that collectively cover the entire coastal region. For all projects, the Planning Tool made use of standardized estimates developed for the Master Plan of the costs and effects on the coast of the preexisting project concepts under consideration. Risk-reduction projects considered by the Master Plan include both structural projects, such as new or improved levees, and nonstructural projects, such as those related to raising the elevation of residences. Restoration projects considered by the Master Plan include various types of projects that have some effect on land building.

In total, the Master Plan considered

- 33 structural risk-reduction project concepts, such as new levee alignments and raising or improving existing levees
- 116 nonstructural risk-reduction projects (parish or subparish programs to elevate residential structures to a specific height above the FEMA base flood elevation, flood-proof some residences and commercial properties, or buy out residential and commercial properties facing extreme flood risk)
- 248 restoration project concepts of nine different types (bank stabilization, barrier island restoration, channel realignment, sediment diversion, hydrologic restoration, marsh creation, oyster barrier reef, ridge restoration, and shoreline protection).

Figures 3.1 and 3.2 show the spatial distribution of these projects across the coast.

Figure 3.1
Locations of Risk-Reduction Projects Evaluated by the Planning Tool

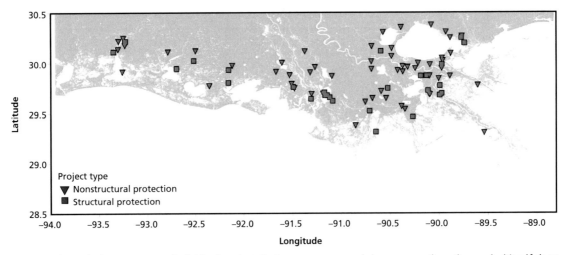

NOTE: Each symbol represents an individual project that may cover a much larger area than the symbol itself does, such as an entire parish.

RAND TR1266–3.1

Figure 3.2
Locations of Restoration Projects Evaluated by the Planning Tool

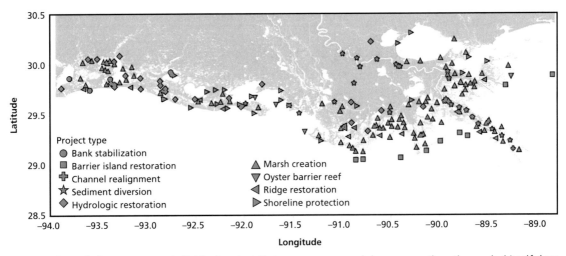

NOTE: Each symbol represents an individual project that may cover a much larger area than the symbol itself does.

RAND TR1266–3.2

Project Costs and Duration of Implementation

CPRA used a standardized approach to develop estimates of project costs and the duration of implementation for three project phases: engineering and design, construction, and O&M. CPRA (2012c, Appendix A) provides more detail about the information and methods used to develop these estimates. Despite the use of consistent methodologies, significant uncertainty exists about project costs and duration. The Planning Tool is designed to evaluate how alternatives would differ under different assumptions, such as the cost of an individual project. This capability was not exercised for the Master Plan; a single, 50-year cost was developed for each project. Table 3.1 summarizes the ranges of these costs for each project type. All project costs were provided in constant 2010 dollars.

Conflicts Among Projects

Lists of incompatible and conflicting projects were developed. At most, only one project from each list could be included in any alternative. Examples of such conflicts include multiple specifications for the same system of barriers or levees (e.g., the Morganza-to-the-Gulf levee system), two nonstructural options for a specific community, and multiple discharge regimes for a diversion project location.

Additional Project Attribute Information

CPRA also compiled information relating to a project's sediment requirement or its use of the flow of the Mississippi or Atchafalaya River (see CPRA, 2012c, Appendix B, for more information). The Master Plan team estimated the amount of sediment required to construct each project and assigned each project to a specific "borrow site" from which sediment could be taken. Sediment-diversion projects were assigned to a specific reach of the Mississippi or Atchafalaya River, and an estimate of the amount of river flow diverted by each project was developed.

Modeling Project Effects

The predictive modeling teams modeled the risk-reduction effects of a large set of risk-reduction projects and the effects on land and other ecosystem-service metrics for a large set of restoration projects for the two environmental scenarios. This information was used by the Planning Tool to compare projects and assemble alternatives.

Table 3.1
Range of Individual Project Costs for Master Plan Projects, by Type

Project Type	Number of Projects	Project Costs (million 2010 $s)		
		Low	Median	High
Nonstructural protection	116	0.2	231	10,512
Structural protection	34	56.2	817	3,964
Bank stabilization	6	11.9	48	169
Barrier island restoration	9	49.1	343	1,830
Channel realignment	9	73.5	4,371	5,583
Sediment diversion	40	16.7	223	13,754
Hydrologic restoration	25	0.6	16	681
Marsh creation	110	32.2	1,454	9,882
Oyster barrier reef	5	18.7	22	171
Ridge restoration	16	1.7	33	70
Shoreline protection	28	4.3	86	1,121

Flood Risk-Reduction Effects

Risk-reduction estimates for individual projects were made for conditions expected to prevail in 50 years. Risk-reduction projects were assumed to obtain their full effect upon completion of construction and then maintain that effect throughout the planning time horizon. Structural projects were assumed to provide no benefits until construction is completed. Nonstructural projects provide benefits proportional to the percentage of completion. A single set of demographic and land-use assumptions was used for the development of the Master Plan (see CPRA, 2012c, Appendix D25). The risk estimates for the Planning Tool are summarized for the communities and regions across the coast in Figure 3.3.

Restoration Project Effects

The effects of projects on ecosystem-service metrics listed earlier were estimated for years 5, 10, 20, 30, 40, and 50 in the implementation of the Master Plan.[1] Effects were assumed to have linear trends within the time intervals. The ecosystem-service metrics, which are developed on a more disaggregated spatial scale, were summarized over each of the 12 regions of the coast for use with the Planning Tool. The 12 regions are shown in Figure 3.4.

Comparison of Individual Projects

The Planning Tool used the project information described earlier to compare each risk-reduction and restoration project to all other projects in the same project category. Comparing individual projects provides a consistent, objective basis for understanding why some projects were included or excluded from alternatives in subsequent steps of the master planning process.

Figure 3.3
Map of the Communities and Regions That Summarize Risk Outcomes

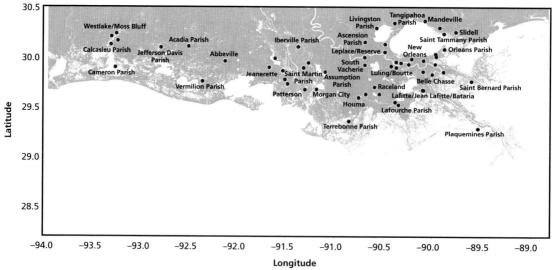

NOTE: Only some communities and regions are labeled.

---

[1] Nature-based tourism data were provided for years 10, 20, 30, 40, and 50. Land-area data were also provided for year 0.

Figure 3.4
Map of the Regions That Summarize Ecosystem-Service Metrics

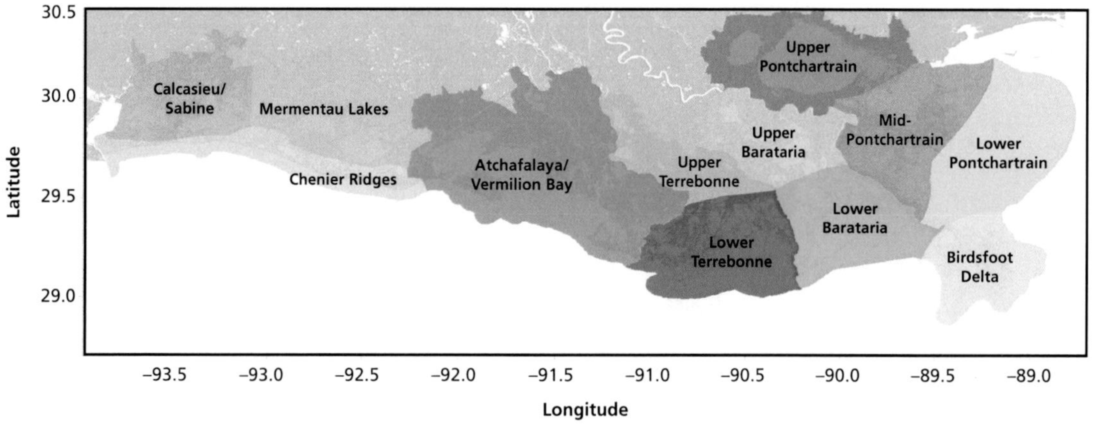

The comparison of projects is based on reduction in EAD for risk-reduction projects and building of land for restoration projects. The Planning Tool also compares projects based on the amount of residual damage at the interval corresponding to each community's risk target, a project's effect on different ecosystem-service metrics, and the other decision criteria. Nearly all these metrics are assumed to be additive, meaning that the total score for two projects for any metric can be represented as the sum of the individual projects' scores. The decision criterion for navigation is the one exception to the additivity assumption.

Project Effects on Risk Reduction

The Planning Tool calculates the long-term reduction in risk due to each project using estimates of EAD from the predictive models for FWOA and FWP conditions, as illustrated in Equation 3.1.

Let p be a member of the set of risk-reduction projects. Let c be a member of the set of communities for which residual damages are calculated.

$$\text{long-term reduction in EAD}_p = \frac{\sum_c \left(\text{EAD}_c^{FWOA} - \text{EAD}_{p,c}^{FWP} \right)}{\sum_c \text{EAD}_c^{FWOA}}.$$

3.1

FWOA refers to FWOA conditions, and *FWP* refers to FWP conditions. Near-term reduction in EAD is calculated in a similar manner to its long-term counterpart. The only difference is that the near-term project effects are measured at 20 years rather than at 50 years.

In addition to calculating risk reduction as measured by EAD, three other measures of risk reduction representing achievement of risk reduction targets are also calculated for their respective recurrence intervals. These three measures are calculated as the ratio of a project's effect on residual damage divided by the coast-wide FWOA level of residual damage for a given flood recurrence interval. This is shown in Equation 3.2.

Let f be a member of the set of recurrence intervals, and let c_f be a member of the set of communities targeted for the recurrence interval f. Let p be a member of the set of risk-reduction projects.

$$\text{risk target achievement}_{p,f} = \frac{\sum_{c_f}\left(\text{residual damage}_{f,c_f}^{FWOA} - \text{residual damage}_{p,f,c_f}^{FWP}\right)}{\sum_{c_f}\text{residual damage}_{f,c_f}^{FWOA}}.$$

$$3.2$$

The numerator of the ratio represents the reduction in residual damage due to the project across the set of communities targeted for risk reduction at the given recurrence interval.[2] The denominator of the ratio represents the total level of residual damage under FWOA conditions across the set of communities targeted for risk reduction at the given recurrence interval.

In both near-term and long-term calculations, the Planning Tool accounts for a project's planning, design, and construction time before its effects on EAD or on residual damage at a specified recurrence interval begin to take effect. For example, if a project requires ten years to plan, design, and construct, then it would have only 40 years postconstruction to affect a long-term metric measured at year 50 (assuming that it started its planning, design, and construction sequence at the beginning of the 50-year time horizon).

Project Effects on Land and Ecosystem-Service Metrics
The Planning Tool calculates changes in land area attributable to a project as the difference in land under FWP and FWOA conditions at year 50. This value is also scaled by dividing changes in land by the coast-wide difference between current and FWOA land area at year 50. For this metric, a ratio equal to 1 would indicate that the amount of land that would otherwise be lost by year 50 without any projects would be recreated as a consequence of the project. See Equation 3.3.

Let r be a member of the set of all regions for which ecosystem-service metrics are calculated. Let p be a member of the set of candidate restoration projects. For each project p,

progress toward maintaining current coast-wide land area $_p$

$$= \frac{\sum_r\left(\text{land area}_{p,r}^{FWP} - \text{land area}_r^{FWOA}\right)}{\sum_r\left(\text{land area}_r^{current} - \text{land area}_r^{FWOA}\right)}.$$

$$3.3$$

Similar to how it handles risk-reduction metrics, the Planning Tool accounts for a project's planning, design, and construction time before its effects on land area begin to take effect.

The effect that a restoration project will have on each of the 15 ecosystem-service metrics is calculated as the difference between the FWP metric value and the FWOA metric value.[3] This calculation can be made at the region level or summed across the regions to represent a coast-level effect. See Equation 3.4.

---

[2] For this analysis, risk reduction was evaluated in 56 communities across the coast. Each community was assigned to a single target level of protection corresponding to a storm flood recurrence interval (e.g., 50-, 100-, or 500-year recurrence).

[3] Project-level data for the agriculture ecosystem-service metric were not available. In addition, brown shrimp and white shrimp were evaluated separately and evaluated combined for use as three separate metrics (brown shrimp, white shrimp, and all shrimp).

Let p be a member of the set of restoration projects and ES_m be the score for ecosystem-service metric m.

$$\text{change in ecosystem metric}_{p,m} = ES_{p,m}^{FWP} - ES_{m}^{FWOA}.$$

3.4

Similar to how it handles risk-reduction metrics and land area metrics, the Planning Tool accounts for a project's planning, design, and construction time before its effects on ecosystem-service metrics begin to take effect.

Project Effects Relative to Other Decision Criteria

For some decision criteria listed in Table 2.2 in Chapter Two, the Planning Tool assimilated external calculations of each project's decision-criteria score. For other decision criteria, the Planning Tool calculated decision-criteria scores for each project using information from the predictive models. This information is helpful for comparing projects relative to the Master Plan objectives.

Cost-Effectiveness

To compare and rank projects, each project score (e.g., risk-reduction metrics, land-building metrics, ecosystem-service metrics, decision criteria) is also scaled by the project's estimated 50-year cost (in 2010 constant dollars and including O&M for the period of time from completion of construction to the end of the 50-year planning time horizon). This provides a measure of a project's cost-effectiveness for a given metric or decision criterion.

Let τ be a member of the set of metrics and decision criteria for which a project is scored. For each project p, the measure of cost-effectiveness for τ is calculated as shown in Equation 3.5.

$$\text{cost-effectiveness measure}_{p,\tau} = \frac{\left(\text{metric or decision-criterion score}\right)_{p,\tau}}{\text{cost}_p}.$$

3.5

The Planning Tool can rank individual projects by their cost-effectiveness score for each metric, decision criterion, and scenario (moderate and less optimistic). To emphasize broad differences among the projects, projects that score highly using the cost-effectiveness measure across many metrics and decision criteria (for both scenarios) are highlighted as highly consistent with the Master Plan objectives. Projects that score highly for only some or none of the metrics and decision criteria are highlighted as sometimes or never consistent with the Master Plan objectives. This information helps explain why the Planning Tool may include or exclude a project in the alternatives.

Formulation of Alternatives

CPRA used the Planning Tool in a four-step interactive process to formulate a range of alternatives to support internal and stakeholder deliberations on projects that could make up the Master Plan. These steps help illustrate the sensitivity of outcomes to different types of alternatives that reflect different policy decisions on funding and time horizon, emphases on objectives, and incorporation of additional expert knowledge:

- *Funding split:* The balance of funding between projects designed primarily for risk reduction and projects designed for coastal restoration
- *Near-term versus long-term benefits:* The balance between achieving benefits in 20 years and achieving benefits in 50 years
- *Emphases on decision criteria and other metrics:* The effect of different constraints on decision criteria, such as use of natural processes, and metrics, such as shrimp, on the Planning Tool's selection of projects for an alternative
- *Incorporation of additional expert knowledge:* The effect of additional rules governing the formulation of an alternative on the Planning Tool's selection of projects for an alternative.

Chapter Four describes the analysis conducted in each of these steps, provides example results, and documents key decisions made along the way.

Integrated Evaluation of Alternatives

As described earlier, the Planning Tool was used to formulate alternatives based on estimates of the individual effects of projects without considering the effects of restoration projects on risk-reduction and structural risk-reduction projects on the ecosystem. Time limitations for developing the Master Plan precluded formulating alternatives based on modeled assessments of the complete alternatives. However, after the final Master Plan alternative was developed and described in the Master Plan, the Master Plan team conducted an integrated evaluation of the draft and final Master Plan. The modeling teams used the predictive models to evaluate the combined effects of all the projects included in the draft and final Master Plan on the coast. This was accomplished by simulating coastal outcomes over time with all the Master Plan projects in place concurrently, according to their implementation schedules. This approach uses the predictive models, and the physical and statistical outcomes generated by them, to estimate how the projects interact with one another rather than assuming that the effects of the individual projects are additive. Fischbach et al. (forthcoming) describe this process for the risk outcomes. The Planning Tool was then used to compare outcomes from the integrated assessment of the alternatives with outcomes calculated by the Planning Tool using individual project-level data and the additive assumptions described earlier. This analysis helped establish when the underlying additive assumption used in the Planning Tool was valid.

Evaluation of Selected Alternatives Using Predictive Models Under Uncertainty
When formulating alternatives, the Planning Tool estimates the effects of implementing individual projects in year 1 for the first implementation period, in year 21 for the second implementation period, and in year 41 for the third implementation period. For the integrated analysis, the predictive models specify that all projects to be implemented in the first period would begin at the beginning of the simulation. Projects to be implemented in either the second or third periods would be specified to begin at year 25 of the simulation. This adjustment of the start times for projects between the alternative-formulation analysis and the integrated alternative evaluation was required because of the high computational requirements of the predictive models but leads to some substantive differences in the results, as described in "Post–Master Plan Analysis" in Chapter Four.

The restoration projects were first evaluated together to determine the evolution of land loss over the 50-year time horizon. The landscape modified by the restoration projects was then used by the storm-surge and damage models to estimate year 50 risk reduction after the risk-reduction projects are implemented. This analysis was performed for the scenarios described earlier plus an additional, more pessimistic scenario.

This approach relaxes three key assumptions based on individual project effect modeling used in the alternative-formulation process: (1) individual project effects are additive, (2) restoration projects do not affect risk, and (3) risk-reduction projects do not affect ecosystem-service metrics.

Comparisons of the Alternatives

The Planning Tool was next used to compare the ways in which the draft and final Master Plan would affect the coast based on the integrated assessment. Special attention was paid to how the alternative as a whole affects future risk, land area, and ecosystem-service metrics, as compared with the estimates developed by the Planning Tool when considering only individual project effects.

Analyses to Develop the Master Plan

CPRA used the Planning Tool and associated analyses to support the development of the draft Master Plan (released in January 2012) and the Master Plan (published in March 2012) (CPRA, 2012a, 2012c). Figure 4.1 lists the key analytic steps and outcomes for these analyses. CPRA and RAND shared many of these results with Louisiana stakeholders and decision-makers during the fall and winter of 2011–2012 using the Planning Tool's interactive Tableau interface (see "Tableau Results Visualizer" in Chapter Two). The Planning Tool analysis was extensive, with significant information generated for each project and alternative. This chapter offers only a snapshot of results intended to illustrate how the Planning Tool was actually used to support the Master Plan.

Compare Individual Projects

CPRA first used the Planning Tool to compare the estimated performance of each risk-reduction project with that of the other risk-reduction projects and to compare the estimated performance of each restoration project with that of the other restoration projects. These com-

Figure 4.1
Planning Tool Analysis and Outcomes for the Master Plan

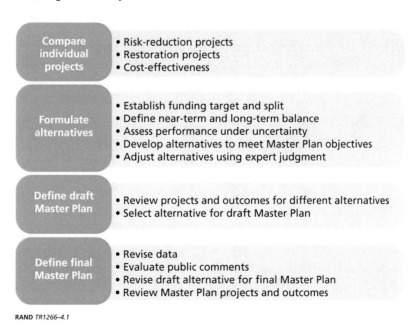

Compare individual projects
- Risk-reduction projects
- Restoration projects
- Cost-effectiveness

Formulate alternatives
- Establish funding target and split
- Define near-term and long-term balance
- Assess performance under uncertainty
- Develop alternatives to meet Master Plan objectives
- Adjust alternatives using expert judgment

Define draft Master Plan
- Review projects and outcomes for different alternatives
- Select alternative for draft Master Plan

Define final Master Plan
- Revise data
- Evaluate public comments
- Revise draft alternative for final Master Plan
- Review Master Plan projects and outcomes

RAND *TR1266–4.1*

parisons allowed CPRA to gain insights into the range of possible effects relative to the Master Plan goals and objectives. By dividing these estimated effects by total cost, CPRA was able to compare the cost-effectiveness of each project.

Risk-reduction projects were compared with one another based on year 50 coast-wide reduction in EAD and reduction in residual damage associated with 50-, 100-, and 500-year flood event recurrence intervals. Table 4.1 shows the estimated range of risk changes due to structural and nonstructural projects for both the moderate and less optimistic scenarios. Structural risk reduction projects have the potential to reduce damages much more significantly than nonstructural risk reduction projects, as shown by the maximum values in Table 4.1. The table also shows that some structural projects that are intended to decrease risk actually increase risk overall by inducing flooding in adjacent areas or trapping overtopped surge or wave water for some storm events. Nonstructural projects, in contrast, never increase risk.

Restoration projects were compared with one another based on their ability to maintain or build land in the near term (at year 20) and in the long term (at year 50). Table 4.2 shows the range in land building for each restoration project type in the near and long term for the moderate and less optimistic scenarios. The projects that have the potential to increase land the most are channel realignment, diversion, and marsh-creation projects. The range of land building for other project types is limited.

To compare projects of different sizes, the risk-reduction and land-building effects are divided by total project cost to yield a cost-effectiveness score for each project. For risk-reduction projects, cost-effectiveness is expressed in terms of reduction in EAD in year 50 per dollar of investment (dollar of EAD reduction divided by dollar of project cost). For restoration projects, cost-effectiveness scores are calculated in terms of near-term land built (year 20) per dollar of investment (square miles of land in year 20 divided by dollars of project cost) and long-term land built (year 50) per dollar of investment (square miles of land in year 50 divided by dollars of project cost).

Figure 4.2 shows the cost-effectiveness scores for the 20 most cost-effective risk-reduction projects. Many of the most cost-effective projects are structural risk-reduction projects (e.g., Greater New Orleans High Level Project), but some nonstructural projects (e.g., that for Saint James Parish) are also cost-effective compared with the structural projects. Figures 4.3 and 4.4 show the cost-effectiveness scores, based on year 50 land changes, for the ten most cost-effective sediment-diversion projects and ten most cost-effective marsh-creation projects, respectively. Note that the horizontal scales are different. Diversion projects are significantly more cost-effective in year 50 than the marsh-creation projects. However, in the near term (year 20), diversion projects have very low cost-effectiveness scores (in some cases, even negative) because diversion projects build land slowly over time.

Formulate Alternatives

CPRA next used the Planning Tool to iteratively develop and evaluate a large set of alternatives. For each iteration, the RAND team used the Planning Tool to formulate different alternatives. These results were provided to CPRA through an interactive, computer-based interface. CPRA then reviewed the analysis, shared selected results with its stakeholders, and provided the RAND team with revised specifications for additional alternatives.

Table 4.1
Range of Risk Reduction for Each Risk-Reduction Project Type, by Environmental Scenario

Project Type	Scenario	Reduction in EAD ($ millions)		Reduction in 50-Year Residual Damage ($ millions)		Reduction in 100-Year Residual Damage ($ millions)		Reduction in 500-Year Residual Damage ($ millions)	
		Minimum	Maximum	Minimum	Maximum	Minimum	Maximum	Minimum	Maximum
Structural	Moderate	−647	2,115	−26,965	32,928	−27,445	48,051	−56,691	456,958
	Less optimistic	−765	10,443	−26,947	453,673	−67,116	484,840	−16,235	161,493
Nonstructural	Moderate	0	221	0	5,114	0	8,534	0	43,740
	Less optimistic	0	1,214	0	52,934	0	48,755	0	39,458

Table 4.2
Range of Net Land-Area Change for Each Restoration Project Type, by Environmental Scenario

Project Type	Number of Projects	Scenario	Near-Term Land (sq. mi.)		Long-Term Land (sq. mi.)	
			Minimum	Maximum	Minimum	Maximum
Bank stabilization	9	Moderate	0	1	0	1
		Less optimistic	0	3	0	1
Barrier island restoration	9	Moderate	0	15	0	15
		Less optimistic	0	14	0	15
Channel realignment	9	Moderate	−3	10	0	154
		Less optimistic	−5	11	0	125
Sediment diversion	40	Moderate	−11	37	−2	37
		Less optimistic	−20	48	−15	138
Hydrologic restoration	25	Moderate	−2	10	−3	10
		Less optimistic	−9	10	−24	45
Marsh creation	108	Moderate	0	0	0	52
		Less optimistic	0	0	−4	52
Oyster barrier reef	5	Moderate	0	1	0	1
		Less optimistic	4	8	5	18
Ridge restoration	16	Moderate	0	1	0	1
		Less optimistic	0	1	0	1
Shoreline protection	25	Moderate	0	2	0	2
		Less optimistic	0	4	0	2

This iterative process helped inform CPRA decisions about allocating funding between risk-reduction and restoration projects and the relative emphasis to place on near-term versus long-term benefits. The analysis also helped to home in on a draft plan and then a final plan by showing how different alternatives might achieve different planning objectives.

Establish the Funding Target and Funding Split

The RAND team and CPRA used the Planning Tool to develop alternatives that maximized risk reduction and land building for different funding scenarios and allocations of funding between risk-reduction projects and restoration projects—the funding split. The Planning Tool was then used to show how risk-reduction and land-building achievement differed across the alternatives.

For this analysis, the RAND team evaluated the two funding scenarios based on different projections of funding streams (see "Mutually Exclusive Project and Project Inclusion or Exclusion Constraints" in Chapter Two)—$20 billion (low funding) and $50 billion (high funding). The RAND team also developed several other funding scenarios based on uniform annual funding levels totaling $30 billion, $40 billion, and $100 billion. CPRA specified that the Planning Tool consider funding splits ranging between 30 percent risk-reduction projects

Figure 4.2
Cost-Effectiveness Scores for the 20 Most Cost-Effective Risk-Reduction Projects

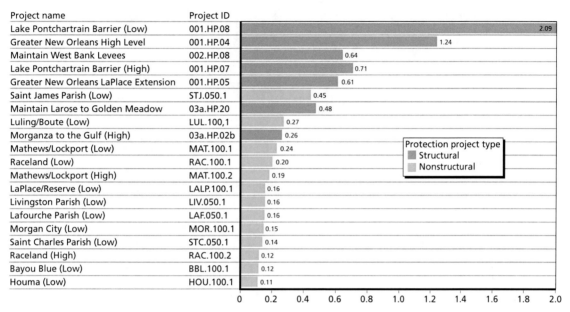

Project name	Project ID	
Lake Pontchartrain Barrier (Low)	001.HP.08	2.09
Greater New Orleans High Level	001.HP.04	1.24
Maintain West Bank Levees	002.HP.08	0.64
Lake Pontchartrain Barrier (High)	001.HP.07	0.71
Greater New Orleans LaPlace Extension	001.HP.05	0.61
Saint James Parish (Low)	STJ.050.1	0.45
Maintain Larose to Golden Meadow	03a.HP.20	0.48
Luling/Boute (Low)	LUL.100,1	0.27
Morganza to the Gulf (High)	03a.HP.02b	0.26
Mathews/Lockport (Low)	MAT.100.1	0.24
Raceland (Low)	RAC.100.1	0.20
Mathews/Lockport (High)	MAT.100.2	0.19
LaPlace/Reserve (Low)	LALP.100.1	0.16
Livingston Parish (Low)	LIV.050.1	0.16
Lafourche Parish (Low)	LAF.050.1	0.16
Morgan City (Low)	MOR.100.1	0.15
Saint Charles Parish (Low)	STC.050.1	0.14
Raceland (High)	RAC.100.2	0.12
Bayou Blue (Low)	BBL.100.1	0.12
Houma (Low)	HOU.100.1	0.11

Protection project type: Structural / Nonstructural

Risk reduction project cost-effectiveness score
(dollars EAD reduction divided by dollars project cost)

NOTE: A score is calculated by dividing the EAD in year 50 by the total cost of a project. Results shown are for the moderate scenario. The horizontal scale is truncated at 2.0.
RAND TR1266–4.2

Figure 4.3
Cost-Effectiveness Scores for the Ten Most Cost-Effective Diversion Projects

Project name	Project ID	
Upper Breton Diversion (5,000 cfs)	001.DI.14	1.210
Fort St. Phillip Diversion (5,000 cfs max capacity)	001.DI.06	0.417
West Pointe a la Hache Diversion (5,000 cfs)	002.DI.06	0.256
Hermitage Diversion (5,000 cfs)	002.DI.18	0.227
Mid-Barataria Diversion (5,000 cfs)	002.DI.02	0.209
Spanish Pass Diversion (7,000 cfs)	002.DI.01	0.193
Mid-Barataria Diversion (50,000 cfs–1st increment)	002.DI.03	0.183
Mid-Breton Sound Diversion (5,000 cfs)	001.DI.23	0.180
Upper Breton Diversion (50,000 cfs)	001.DI.15	0.174
Lower Breton Diversion (5,000 cfs)	001.DI.01	0.160

Restoration project cost-effectiveness score
(square miles of land divided by millions of dollars of project costs)

NOTE: Scores are calculated by dividing the net land that the project creates by year 50, in square miles, by the total cost of the project (square miles of land in year 50 divided by the project cost in millions of dollars). Results shown are for the moderate scenario. The horizontal scale is different from that in Figure 4.4.
RAND TR1266–4.3

Figure 4.4
Cost-Effectiveness Scores for the Ten Most Cost-Effective Marsh-Creation Projects

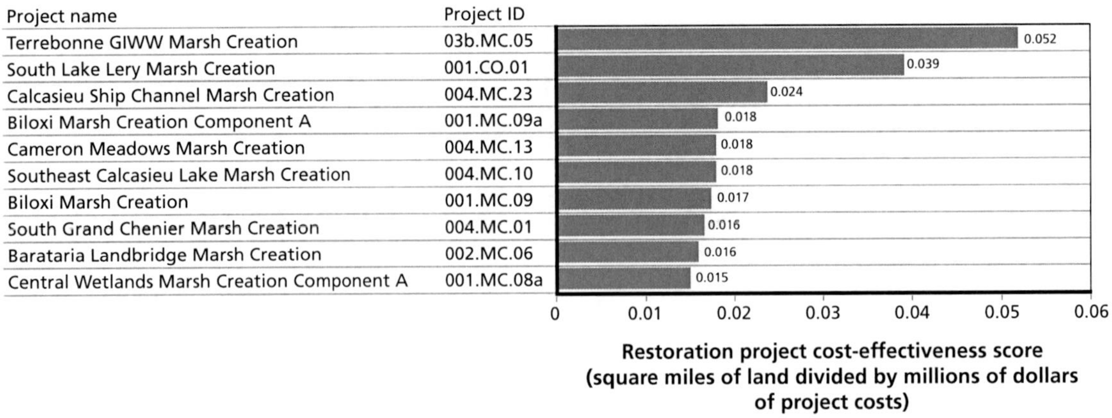

Project name	Project ID	
Terrebonne GIWW Marsh Creation	03b.MC.05	0.052
South Lake Lery Marsh Creation	001.CO.01	0.039
Calcasieu Ship Channel Marsh Creation	004.MC.23	0.024
Biloxi Marsh Creation Component A	001.MC.09a	0.018
Cameron Meadows Marsh Creation	004.MC.13	0.018
Southeast Calcasieu Lake Marsh Creation	004.MC.10	0.018
Biloxi Marsh Creation	001.MC.09	0.017
South Grand Chenier Marsh Creation	004.MC.01	0.016
Barataria Landbridge Marsh Creation	002.MC.06	0.016
Central Wetlands Marsh Creation Component A	001.MC.08a	0.015

Restoration project cost-effectiveness score
(square miles of land divided by millions of dollars
of project costs)

NOTE: GIWW = Gulf Intracoastal Waterway. Scores are calculated by dividing the net land that the project creates by year 50, in square miles, by the total cost of the project (square miles of land in year 50 divided by the project cost in millions of dollars). Results shown are for the moderate scenario. The horizontal scale is different from that in Figure 4.3.
RAND TR1266-4.4

to 70 percent restoration projects (30/70) and 70 percent risk-reduction projects to 30 percent restoration projects (70/30). The Planning Tool was used to generate a total of 25 alternatives, one for each of the 25 possible combinations of five total funding amounts and five funding splits:

- five 50-year funding amounts, in 2010 dollars
  - $20 billion (low funding)
  - $30 billion
  - $40 billion
  - $50 billion (high funding)
  - $100 billion
- five funding splits (risk reduction project percentage/restoration project percentage)
  - 30/70
  - 40/60
  - 50/50
  - 60/40
  - 70/30.

Results were generated under different specifications of how to balance near-term and long-term risk reduction and land building and for the two environmental scenarios. Figure 4.5 shows results for long-term risk reduction and land building when equally weighting near-term and long-term results for the moderate scenario. As expected, increased total funding leads to higher risk reduction and a greater amount of land built (shifts to the upper right). Similarly, increasing funding for risk reduction at the expense of land building increased long-term risk reduction and decreased long-term land built. Results with different emphases on near-term and long-term benefits showed similar patterns.

As funding levels increase, more land can be built, but there are limits to the amount of risk that can be reduced. Specifically, for the $50 billion and $100 billion funding scenarios, improvements in risk reduction appeared to be unattainable beyond 83 percent. This is because risk-reduction benefits are ultimately limited by the expected performance of candidate projects under consideration and no feasible combination of projects is able to eliminate all risk. For land building, however, it is possible that, with enough funding, multiple restoration projects can together increase total coast-wide land area above current levels. This is represented in Figure 4.5 by a symbol being above the 100-percent line for long-term increases in land. It is important to note that, to reach this level of land building, nearly $40 billion of funding would need to be allocated solely to restoration efforts. Data points for more-realistic funding levels all fall below the 100-percent line, indicating that, under more-realistic funding levels, the state cannot achieve a future equal in land area to current conditions unless more than 70 percent of available funding is allocated to restoration projects.

Define the Near-Term and Long-Term Balance

The RAND team next formulated alternatives that varied emphasis on near-term and long-term outcomes. The Planning Tool calculated ten alternatives that incrementally varied this balance between 0 percent near term/100 percent long term and 90 percent near term/10 percent long term.[1] Each alternative was based on a total $50 billion, 50-year budget, split equally

Figure 4.5
Long-Term Risk Reduction and Long-Term Land Building for Different Funding Splits and Total Funding Level

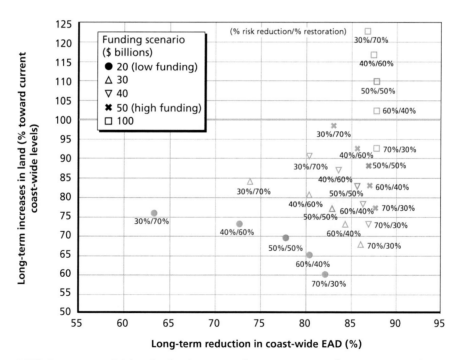

NOTE: Percentage of risk reduction is presented as a percentage of FWOA EAD. Land building is presented as a percentage of land lost under FWOA conditions. Long-term results are those for year 50. Symbols indicate different funding scenarios. Labels indicate different funding splits. Results are for the moderate scenario. Results for a 50/50 split are colored red.

RAND *TR1266–4.5*

[1] The results for 100 percent near term/0 percent long term balance are not calculated because the Planning Tool assigns no benefit to selecting any projects after year 20 in this case and thus does not expend the full 50-year budget.

CPRA Decisions

CPRA concluded that a balanced approach provided the appropriate funding split between risk-reduction and land-building results for both near-term and long-term case, particularly for the $50 billion funding case, whose results are marked in Figure 4.5 by x's. In the $50 billion funding scenario, when the funding allocation for risk reduction is greater than 50 percent, the results show significant diminishing returns in risk reduction for both near- and long-term results. A strictly equal allocation of funding between risk reduction and restoration projects was used to develop the draft Master Plan but then relaxed somewhat for the final Master Plan.

This analysis showed that $20 billion was inadequate to meet the objectives of the Master Plan; securing $50 billion would provide the necessary funding to meet these objectives. CPRA chose to focus the remaining analysis on the $50 billion funding scenario.

---

between risk-reduction and restoration projects, consistent with the decisions on funding splits and amounts taken by CPRA in the preceding step.[2]

Figure 4.6 shows the structural risk-reduction projects selected for each of the ten alternatives that vary the near-term/long-term balance. The results show that changing the relative balance between near-term and long-term risk-reduction benefits does not significantly change the set of projects selected for implementation, although it does sometimes alter the time period for which a selected project has been chosen for implementation.[3] This result is due to the fact that, once risk-reduction benefits are achieved, they persist for the rest of the 50-year time period. Therefore, as long as some weight is placed on near-term benefits, the Planning Tool will maximize near-term benefits through investment in the first period and then maximize the additional risk reduction through implementing projects in the second and third periods.[4]

Restoration projects each provide different patterns of benefits over time. Some projects, such as marsh-creation projects, provide the most benefits immediately upon completion. Other projects, such as sediment diversions, build land slowly over time. Changing the relative balance of near- and long-term land building thus changes the mix of the specific types of restoration projects included in an alternative.

Figure 4.7 illustrates how changing the balance between near-term and long-term land building leads to different outcomes over time. For example, specifying 100 percent long-term land building (with no explicit emphasis placed on near-term land building) leads to the most land being built in 50 years but results in less land being built in ten, 20, and 30 years.

---

[2] Note that each project has specific timelines for engineering, design, and construction, which are accounted for in the Planning Tool. Therefore, projects do not produce outcomes immediately (i.e., first year of implementation period).

[3] Differences in the set of projects selected for implementation under different combinations of near-term and long-term weightings are based on the available funding prior to the time period (near or long term) and how that affects the sequence of projects that can be selected early enough to affect risk reduction at the time period of interest. When long-term benefits receive 80 percent weight or more, the Planning Tool forgoes the Amelia Levee Improvements 3E project in the first period and its near-term benefit, and instead the Planning Tool selects the Amelia Levee Improvements 1E project for the third period because this still provides long-term (year 50) benefits that are now the primary focus with 80 percent of the weight. This frees up funds for additional nonstructural projects to be selected in the first and second periods (not shown), thus increasing overall long-term risk reduction but slightly lowering near-term risk reduction.

[4] The Planning Tool optimization algorithm may delay some projects with high O&M costs to later in the 50-year period if that enables it to implement additional projects because of O&M cost savings.

Figure 4.6
Structural Risk-Reduction Projects Selected for Alternatives with Different Balances Between Near-Term and Long-Term Benefits

Project Name	Near Term/Long Term (%)									
	90/10	80/20	70/30	60/40	50/50	40/60	30/70	20/80	10/90	0/100
Abbeville and Vicinity	▪	▪	▪	▪	▪	▪	▪	▪	▪	▪
Amelia Levee Improvements (1E)								▪	▪	▪
Amelia Levee Improvements (3E)	▪	▪	▪	▪	▪	▪	▪			
Berwick to Wax Lake	▪	▪	▪	▪	▪	▪	▪	▪	▪	▪
Franklin and Vicinity	▪	▪	▪	▪	▪	▪	▪	▪	▪	▪
Greater New Orleans High Level	▪	▪	▪	▪	▪	▪	▪	▪	▪	▪
Greater New Orleans LaPlace Extension	▪	▪	▪	▪	▪	▪	▪	▪	▪	▪
Iberia/Vermilion Upland Levee	▪	▪	▪	▪	▪	▪	▪	▪	▪	▪
Lafitte Ring Levee	▪	▪	▪	▪	▪	▪	▪	▪	▪	▪
Lake Pontchartrain Barrier (Low)	▪	▪	▪	▪	▪	▪	▪	▪	▪	▪
Maintain Larose to Golden Meadow	▪	▪	▪	▪	▪	▪	▪	▪	▪	▪
Maintain West Bank Levees	▪	▪	▪	▪	▪	▪	▪	▪	▪	▪
Morgan City Back Levee	▪	▪	▪	▪	▪	▪	▪	▪	▪	▪
Morganza to the Gulf (High)	▪	▪	▪	▪	▪	▪	▪	▪	▪	▪
Southwest GIWW (Med)	▪	▪	▪	▪	▪	▪	▪	▪	▪	▪

Implementation period: ▪ 2012–2031   ▪ 2032–2051   ▪ 2052–2061

RAND *TR1266–4.6*

Figure 4.7
Trends in Coast-Wide Land Area over Time for Moderate Future Conditions

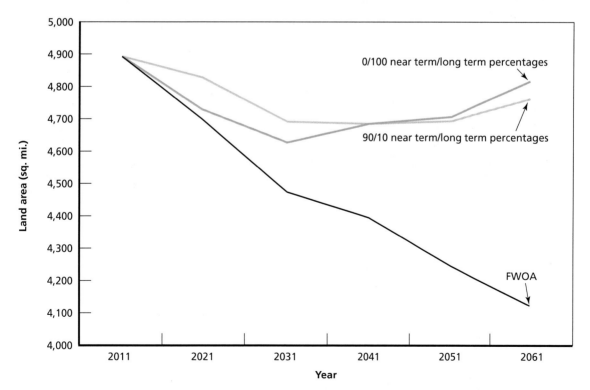

NOTE: The change of slope shown between 2032 (year 20) and 2042 (year 30) on the FWOA curve is a result of landscape updates in the predictive models during this time period. Results are for a moderate environmental scenario and a $50 billion funding scenario.

RAND *TR1266–4.7*

Figure 4.8 summarizes the balance between near-term (year 20) and long-term (year 50) land-building outcomes corresponding to different balances between near- and long-term land building. These results show the expected trade-off curve between near- and long-term coast-wide land building. Interestingly, the ranges in near-term and long-term outcomes are rather restricted: between 175 and 220 square miles for near term and between 640 and 700 square miles for long term, excluding the case in which the Planning Tool maximizes 100 percent near-term benefits.

Changing the balance between near-term and long-term benefits shifts the balance of project expenditures between marsh-creation projects, which provide large benefits in the near term, and diversion projects, which provide large benefits in the long term (Figure 4.9). As the shift toward long-term outcomes is favored, more funds are used for diversion projects rather than for marsh-creation projects, with a sharp change at a balance of 40/60 near-term/long-term land-building outcomes. This change is represented by an increase in expenditures (relative to expenditures when 90 percent weight is placed on near-term land building) on diversion projects as the focus shifts toward increasing the weight on long-term land building. A corresponding decrease in expenditures on marsh-creation projects (relative to expenditures when 90 percent weight is placed on near-term land building) is observed as the focus shifts toward increasing weight on long-term land building.

Figure 4.8
Near-Term and Long-Term Land-Building Results for Different Balances Between Near-Term and Long-Term Outcomes

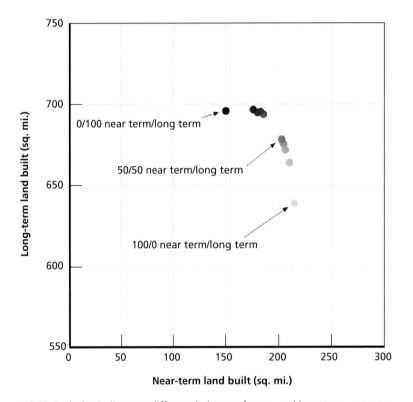

NOTE: Each dot indicates a different balance of near- and long-term outcomes in the Planning Tool objective function. The results reflect a $50 billion total budget using a 50/50 split to allocate funding between restoration and risk-reduction projects.
RAND TR1266-4.8

Figure 4.9
Change in Restoration Project Expenditures, by Project Type, for Different Near-Term/Long-Term Balances

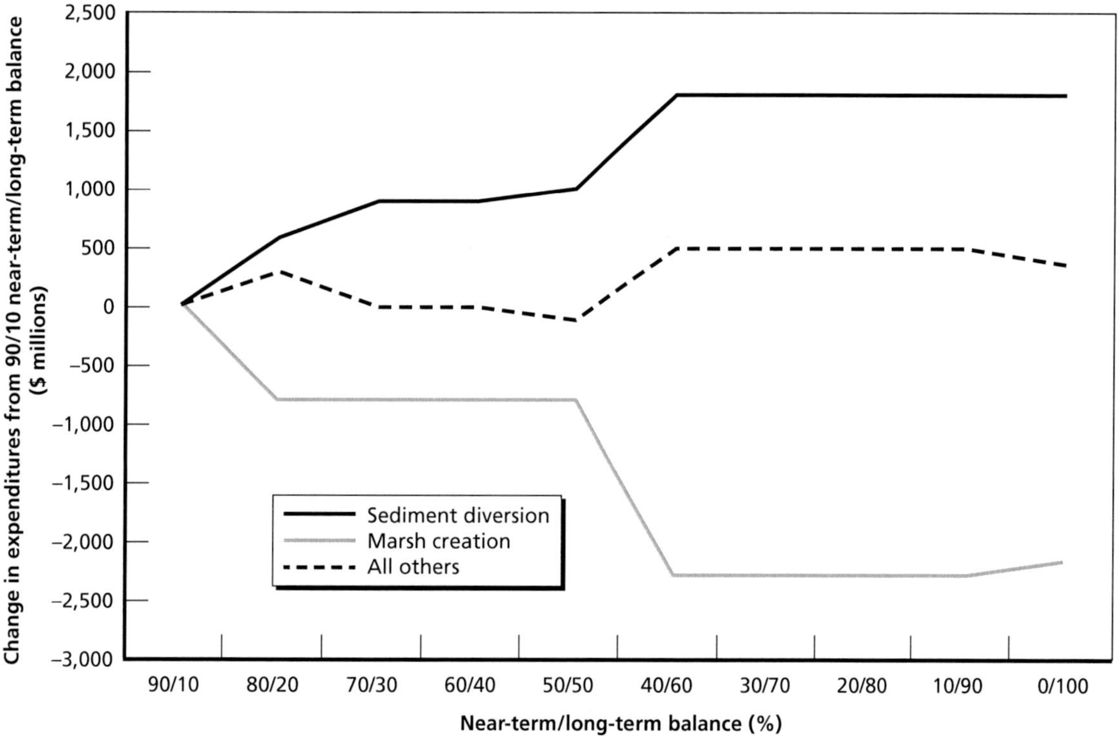

RAND *TR1266–4.9*

```
┌─────────────────────────────────────────────────────────────────────┐
│                         CPRA Decisions                                │
│                                                                       │
│ CPRA decided to balance near-term and long-term risk-reduction and    │
│ land outcomes equally for the remainder of the Planning Tool          │
│ analyses. This 50-percent near-term and 50-percent long-term          │
│ approach balances the need to respond with urgency to the coastal     │
│ crisis while investing in long-term solutions.                        │
└─────────────────────────────────────────────────────────────────────┘
```

Assess Performance Under Uncertainty

The projects selected for an alternative by the Planning Tool differ depending on which environmental scenario is being considered. Those projects that maximize land for the moderate scenario, for example, are different from those projects that maximize land for the less optimistic scenario. As expected, CPRA found that alternatives formulated for the moderate scenario outperform the alternatives formulated for the less optimistic scenario, under moderate scenario conditions. Similarly, alternatives formulated for the less optimistic scenario outperform those formulated for the moderate scenario, under less optimistic scenario conditions.

Figure 4.10 shows how coast-wide risk and land building vary depending on the scenario under which the alternative is formulated and the scenario under which the alternative is evaluated. The alternative formulated under the less optimistic scenario (orange bars) performs slightly less well than the other under moderate scenario conditions (top bars) but performs

Figure 4.10
Comparison of Land Area in Year 50 for Alternatives Developed to Maximize Land Under Either the Moderate or Less Optimistic Scenario

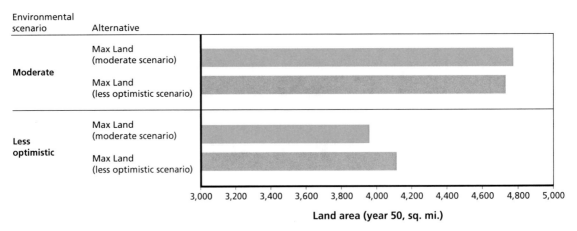

RAND *TR1266–4.10*

CPRA Decisions

CPRA found that restoration projects selected under less optimistic conditions tended to be in the upper end of the estuaries, closer to existing land, than projects close to the Gulf of Mexico. Informed by these results, CPRA chose to base the Master Plan on the projects selected under the less optimistic scenario. This alternative will perform slightly less well than others under moderate conditions but will have greater benefits if conditions similar to the less optimistic scenario come to pass.

---

much better (more than 250 square miles of land in 50 years) under the less optimistic scenario (lower bars).

Develop Alternatives to Meet Master Plan Objectives

The Planning Tool was next used to evaluate how alternatives would change—in terms of projects included and expected outcomes—as metric and decision-criterion constraints were added to emphasize different Master Plan objectives.

To understand how adding constraints may affect the formulation of alternatives, it is useful to recall that the Planning Tool is structured as a "constrained maximization" problem. This means that the Planning Tool, in its simplest form, selects an alternative by identifying a group of projects that fit within available funding, sediment, and river-flow constraints and that maximize the benefits of near- and long-term risk reduction and near- and long-term land building; these are the four terms of the objective function (described in "Basis of the Approach in Decision Theory" in Chapter Two). As such, the addition of other constraints related to particular decision criteria could have one of two effects. If the constraints' inclusion in the Planning Tool has no effect on the selection of projects, then the value of the objective function will be the same as it would have been without the constraint. In contrast, if the inclusion of the constraint in the Planning Tool affects the selection of projects (i.e., the constraint is binding), then the aggregate value of the decision drivers can be only less than it would have been in the absence of the constraint. Adding constraints to the Planning Tool can lead only to alternatives whose aggregate decision-driver value is the same or less than it would be without the constraints.

Important trade-offs between and among planning objectives can be illustrated by exploring how risk reduction and land building change when decision criteria or metric constraints are added. For example, these sensitivity analyses can help answer such questions as these:

- If a requirement were added to increase the use of natural processes, what changes in project selection and land-building benefits would be seen in an alternative that maximizes land building?
- If a requirement were added that coast-wide shrimp habitat needs to be maintained at current levels, what changes in project selection and land-building benefits would be seen in an alternative that maximizes land building?
- If a requirement were added that projects not impede current navigation activity, what changes in project selection and risk-reduction benefits would be seen in an alternative that maximizes risk reduction?

The Planning Tool developed many alternatives that were subjected to constraints on various factors. During this process, CPRA evaluated the effects of all decision criteria and metrics. Table 4.3 lists ten decision criteria and three ecosystem-service metrics that were explored in more depth as part of the alternative-formulation process and the ranges of values used as thresholds for the sensitivity analysis. The range of threshold values used for each decision criterion or metric was developed based on the score that the given criterion or metric received for the alternative formulated when no decision criteria or metric constraints were applied and the score for the given criterion or metric that generated infeasible results.

To analyze results from adding these various constraints, one at a time, to the Planning Tool, four different types of displays were generated from the Planning Tool:

- Risk-reduction graphs: These graphs show how long-term reduction in EAD in year 50 varies under different constraints on decision criteria.
- Land-building graphs: These graphs show how long-term coast-wide land area (in year 50) varies under different constraints on decision criteria and metric outcomes.
- Project inclusion tables: These tables compare the projects included in alternatives with different values on the various constraints.
- Project inclusion frequency tables: These tables show how frequently each project is included across alternatives generated by varying constraints.

Results were analyzed for both the moderate and less optimistic scenarios.

Sensitivity of Risk-Reduction Project Selection to Varying Constraints on Decision Criteria

The Planning Tool was used to test the sensitivity of risk outcomes to various decision criteria. The only decision criterion that made a significant difference on EAD, however, was use of natural processes. Figure 4.11 shows the trade-off between EAD and constraints placed on the *use of natural processes* decision criterion. When no constraint is applied, EAD is reduced as much as possible. Once the *use of natural processes* criterion threshold is greater than a value of −2.0 (i.e., less negative, moving to the right of the graph), progress toward reducing EAD begins to

Table 4.3
Decision Criteria and Metrics Constrained as Part of the Master Plan Sensitivity Analysis

Constraint	Basis for Scoring of Individual Projects	Basis for Scoring the Alternative	Range of Scores Explored in the Analysis
Decision criteria			
Use of natural processes (risk reduction)	Project's use of natural processes (scores range from –1 to 0 per project; see CPRA, 2012c, Appendix B, Attachment B4)	Summation of risk-reduction project scores making up the alternative	≥–3.2 to –1.6
Flood protection of strategic assets (risk reduction)	Additional number of strategic assets that are protected from flooding by the 50-year event (see CPRA, 2012c, Appendix B, Attachment B10)	Percentage of assets at risk to flooding once in 50 years that are protected by the alternative	≥12.8% to 18%
Flood protection of historic properties (risk reduction)	Additional number of historic assets that are protected from flooding by the 50-year event (see CPRA, 2012c, Appendix B, Attachment B8)	Percentage of historic properties at risk to flooding once in 50 years that are protected by the alternative	≥15% to 18%
Distribution of flood risk across socioeconomic groups (risk reduction)	Amount of residual EAD in census tracts that are indicated as impoverished by the 2005–2009 U.S. Census Bureau American Community Survey (see CPRA, 2012c, Appendix B, Attachment B5, and U.S. Census Bureau, 2012)	EAD in year 50 for census tracts indicated as impoverished	≤$150 million to $350 million
Support of navigation (risk reduction)	Project's impact on navigation (scores ranging from –1 to 0 per project; see CPRA, 2012c, Appendix B, Attachment B11)	Alternatives not scored due to nonadditivity of project scores	Alternatives were developed with exclusion of all projects with a score of less than 0 and no exclusion
Sustainability (restoration)	Long-term sustainability of land building as proxied by the long-term trend in land building for projects making up the alternative (see CPRA, 2012c, Appendix B, Attachment B7)	Trend in the sum of land built by each project between 40 and 50 years after implementation, scaled by total amount of land lost by year 50 in FWOA	≥14% to 18%
Use of natural processes (restoration)	Project's use of natural processes (scores range from 0 to 1 per project; see CPRA, 2012c, Appendix B, Attachment B6)	Summation of restoration project scores making up the alternative	≥22 to 30
O&M (restoration)	Amount of O&M costs relative to 50-year costs (see CPRA, 2012c, Appendix B, Attachment B8)	Sum of alternative's O&M expenditures in year 50 compared with average annual funding for the alternative	≤4% to 7%

Table 4.3—Continued

Constraint	Basis for Scoring of Individual Projects	Basis for Scoring the Alternative	Range of Scores Explored in the Analysis
Support of navigation (restoration)	Project's impact on navigation (scores ranging from –1 to 1 per project; see CPRA, 2012c, Appendix B, Attachment B11)	Alternatives not scored due to nonadditivity of project scores	Exclude if <–0.4 to 0; include if >0.3 to 0.1
Additional metrics			
Oyster barrier reef (restoration)	Oyster habitat in units of habitat suitability index times area	Sum of individual project's effect on oyster habitat	≥FWOA to 20% greater than current levels
Shrimp (restoration)	Shrimp habitat in units of habitat suitability index times area	Sum of individual project's effect on shrimp habitat	≥50% of current to current levels
Saltwater fisheries (restoration)	Saltwater fish habitat in units of habitat suitability index times area	Sum of individual project's effect on saltwater fish habitat	≥50% of current to current levels
Critical landforms (additional analysis) (restoration)	Amount of land built by restoration projects associated with LACPR's critical landforms. Projects unassociated with these landforms receive a 0%. Projects associated receive a score equal to the ratio of that project's 50-year land building and the total land building from all projects associated with the landforms (see CPRA, 2012c, Appendix B, Attachment B15)	Total percentage of land built by projects associated with critical landform. A 100% indicates that all projects associated with critical landforms are implemented	≥30% to 100%

decrease significantly.[5] This occurs because the Planning Tool ceases to include major cross-basin levee alignments that score less than –2.0 for the *use of natural processes* decision criterion.

Figure 4.12 shows which risk-reduction projects are included for alternatives that include different constraint levels on the *use of natural processes* decision criterion. Applying a constraint greater than or equal to –2.4 significantly changes the alternative by replacing the extensive Southwest GIWW levee alignment with the much smaller Lake Charles levee alignment along with additional nonstructural protection projects for the western portion of the state. Even tighter constraints on natural processes (values even farther to the right on the graph) eliminate additional levees, including the Lafitte Ring Levee (at ≥–2.0), Lake Charles 500-Year Protection (at ≥–1.6), Morganza to the Gulf (High) (at –0.8 and ≥0), and Lake Pontchartrain Barrier (Low) (at ≥–0.4).[6]

The following is a list of the projects that are always, sometimes, and never included for those cases in which the "use of natural processes" decision-criterion constraint threshold is

[5] The *use of natural processes* decision-criterion score is calculated by summing all the included projects' *use of natural processes* criterion scores (described in CPRA, 2012c, Appendix B, Attachment B6). An alternative's *use of natural processes* criterion score is meaningful only to compare with other alternatives.

[6] There were two variations of alternatives. *High* and *low* refer to different height specifications for the levee alignment.

Figure 4.11
Reduction in Risk Versus the *Use of Natural Processes* Decision Criterion for Ten Alternatives

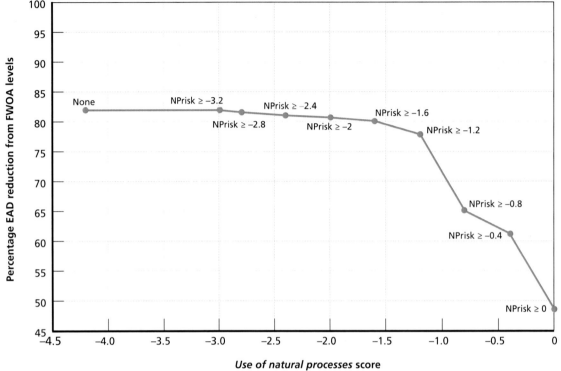

NOTE: Reduction in risk is expressed as a percentage of EAD in FWOA conditions for the moderate environmental scenario. The text labels indicate the constraint specified for the *use of natural processes* decision criterion (e.g., NPrisk ≥ −1.2 specifies a constraint of −1.2).
RAND *TR1266–4.11*

≤−1.6. The structural risk-reduction projects that are always included score well with respect to the use of natural processes.

- always included
  - Berwick to Wax Lake
  - Franklin and Vicinity
  - Greater New Orleans High Level
  - Maintain Larose to Golden Meadow
  - Maintain West Bank Levees
- sometimes included (percentage of cases given in parentheses)
  - Amelia Levee Improvements 2E (70)
  - Amelia Levee Improvements 3E (30)
  - Abbeville and Vicinity (60)
  - Greater New Orleans LaPlace Extension (10)
  - Iberia/Vermilion Upland Levee (10)
  - Lafitte Ring Levee (40)
  - Lake Charles 500-Year Protection (20)
  - Lake Pontchartrain Barrier (Low) (80)
  - Morgan City Back Levee (80)

Figure 4.12
Structural Risk-Reduction Projects Included for Alternatives Generated by Imposing Constraints on the Use of Natural Processes

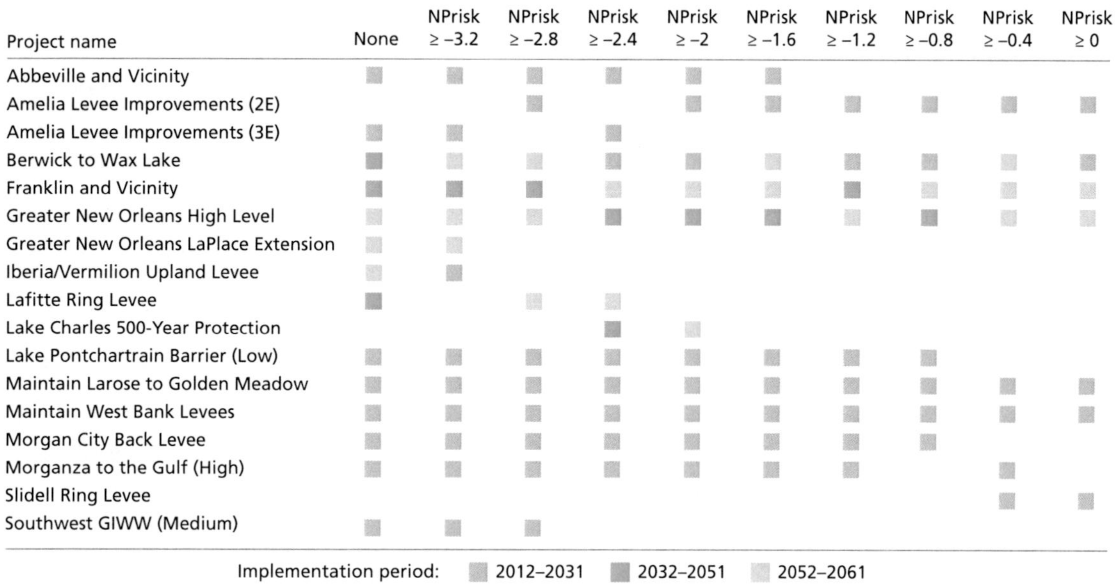

Project name	None	NPrisk ≥ –3.2	NPrisk ≥ –2.8	NPrisk ≥ –2.4	NPrisk ≥ –2	NPrisk ≥ –1.6	NPrisk ≥ –1.2	NPrisk ≥ –0.8	NPrisk ≥ –0.4	NPrisk ≥ 0
Abbeville and Vicinity	■	■	■	■	■	■				
Amelia Levee Improvements (2E)			■		■	■	■	■	■	■
Amelia Levee Improvements (3E)	■	■		■						
Berwick to Wax Lake	■	■	■	■	■	■	■	■		■
Franklin and Vicinity	■	■	■	■	■	■	■	■	■	■
Greater New Orleans High Level	■	■	■	■	■	■	■	■	■	■
Greater New Orleans LaPlace Extension	■	■								
Iberia/Vermilion Upland Levee	■	■								
Lafitte Ring Levee	■		■	■						
Lake Charles 500-Year Protection				■	■					
Lake Pontchartrain Barrier (Low)	■	■	■	■	■	■	■	■		
Maintain Larose to Golden Meadow	■	■	■	■	■	■	■	■	■	■
Maintain West Bank Levees	■	■	■	■	■	■	■	■	■	■
Morgan City Back Levee	■	■	■	■	■	■	■	■		
Morganza to the Gulf (High)	■	■	■	■	■	■	■		■	
Slidell Ring Levee									■	■
Southwest GIWW (Medium)	■	■	■							

Implementation period: ■ 2012–2031   ■ 2032–2051   ■ 2052–2061

RAND *TR1266–4.12*

- Morganza to the Gulf (High) (80)
- Slidell Ring Levee (20)
- Southwest GIWW (Medium) (30)
- never included
  - Amelia Levee Improvements 1E
  - Caernarvon to White Ditch
  - Donaldsonville to the Gulf
  - Gueydan Ring Levee
  - Lake Charles Ring Levee (South)
  - Lake Pontchartrain Barrier (High)
  - Larose to Morgan City
  - Morganza to the Gulf (Low)
  - Oakville to Myrtle Grove
  - Southwest GIWW (High)
  - Southwest GIWW (Low)
  - West Bank High Level.

Sensitivity of Restoration Project Selection to Varying Constraints on Decision Criteria

Coast-wide land area is sensitive to varying decision criteria and metric constraints. For example, Figures 4.13 and 4.14 show the trade-offs between land area built by year 50 and constraints on the minimum outcomes for shrimp and saltwater fisheries, respectively. In both cases, requiring that the habitat suitability score for the ecosystem-service metric be maintained near or at current levels significantly reduces the amount of land that can be built by an alternative by year 50.

Figure 4.13
Trade-Offs Between Change in Land by Year 50 and Shrimp

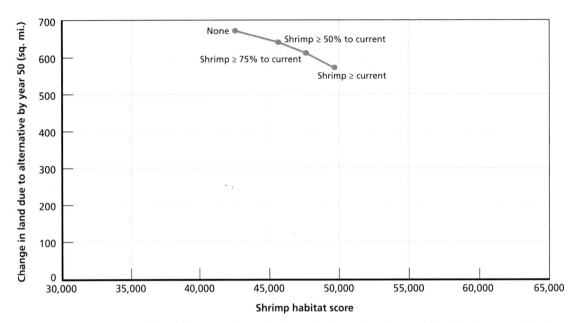

NOTE: Labels indicate the value of the constraint applied in each alternative. Amount of habitat is expressed as the product of a habitat suitability index score (between 0 and 1) and the area of land.
RAND TR1266–4.13

Figure 4.14
Trade-Offs Between Change in Land by Year 50 and Saltwater Fisheries

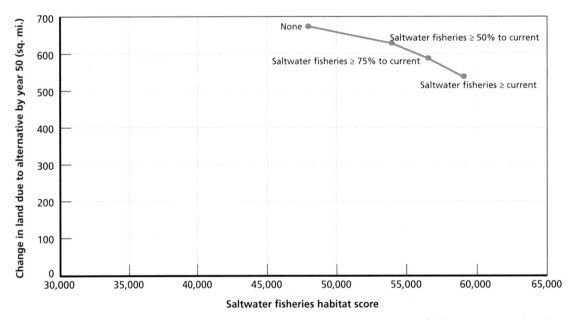

NOTE: Labels indicate the value of the constraint applied in each alternative. Amount of habitat is expressed as the product of a habitat suitability index score (between 0 and 1) and the area of land.
RAND TR1266–4.14

Figure 4.15 shows trade-offs between land area built by year 50 and four other decision-criterion constraints, assuming an alternative that includes no other decision-criterion constraints as a starting point (indicated by "none" in the figures). For the *use of natural processes* decision criterion, it was not feasible to construct an alternative with a score of ≥30. Increasing it from 21, its value for the baseline case, to 28, reduces land area only modestly. Applying a constraint on the *support of navigation* decision criterion, however, has a large impact on land building. Because the *support of navigation* decision-criterion scores are not additive, the horizontal axis for the corresponding graph (upper-right) specifies which project scores result in exclusion or inclusion of a project. If all restoration projects with a negative score for *support of navigation* are excluded ("Exclude Nav < 0" in Figure 4.15), then land built by year 50 declines by more than 50 percent. For both *sustainability* and *O&M*, the effect of constraints on land building in year 50 is modest. Note that alternatives that improved these scores beyond those shown in the figure were infeasible.

As with the risk-reduction alternatives, the Planning Tool reports on the restoration projects that are included under the varying decision-criterion constraints. For example, Figure 4.16 shows how the application of constraints on *support of navigation* reduces the number of large sediment-diversion projects selected. Specifically, excluding projects that have *support of navigation* scores of less than −0.2 (third column) leads the Planning Tool to substitute a 5,000 cfs diversion for a 250,000 cfs diversion at Upper Breton and a 50,000 cfs diversion for a 250,000 cfs diversion at Mid-Barataria.

Table 4.4 summarizes how often the diversion projects are selected over the alternatives that vary decision-criterion constraints. Projects scoring 100 percent were selected regardless of what level of constraint was applied for the given decision criterion or metric. Projects scoring less than 100 percent were not selected under at least one level of the constraint examined. For example, a project with a score of 75 percent for the *use of natural processes* decision criterion was selected under three-quarters of the alternatives that varied the constraint values applied to the *use of natural processes* decision criterion. CPRA used this information to determine which decision criteria could be used to ensure a diversity of approaches for building land.

Table 4.5 lists each preliminary alternative along with the decision criterion and threshold value set in the Planning Tool.

Adjust Alternatives Using Expert Judgment
CPRA next used the Planning Tool to adjust the alternatives based on expert judgment. Each alternative continued to maximize near-term and long-term risk reduction and near-term and long-term land area (using the previously determined equal balance between near- and long-term goals) but also reflected specifications about different projects to include or exclude. These alternatives addressed specific issues raised by stakeholders and helped to explore different mixes of project types, accounting for favorable and unfavorable conditions associated with certain projects not explicitly included in the decision criteria. CPRA (2012c, Appendix A)

CPRA Decisions

CPRA reviewed these sensitivity results by looking at how results for long-term risk reduction and land building changed and how project selection varied under the range of constraints. Using this information, CPRA identified a threshold value for some decision criteria to use in specifying alternatives to be carried forward in the analysis and used to develop *expert-adjusted* alternatives.

Figure 4.15
Trade-Offs Between Land Area Built by Year 50 and Different Decision-Criterion Scores

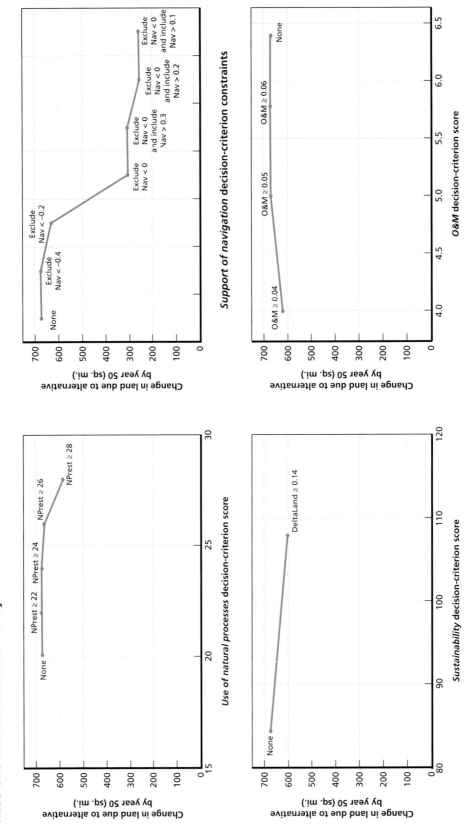

NOTE: Each point presents results corresponding to an alternative with a decision-criterion constraint as labeled. The upper-left figure shows results for the *use of natural processes* decision criterion. The upper-right figure shows results for the *support of navigation* decision criterion. The lower-left figure shows results for the *sustainability* decision criterion. The lower-right figure shows results for the *O&M* decision criterion.

RAND *TR1266-4.15*

Figure 4.16
Sediment Diversion Projects Included in Alternatives That Vary the *Support for Navigation* Criterion

Project Name	Project ID	None	Exclude Nav < -0.4	Exclude Nav < -0.2	Exclude Nav < 0	Exclude Nav < 0 & Include Nav > 0.3	Exclude Nav < 0 & Include Nav > 0.2	Exclude Nav < 0 & Include Nav > 0.1
Atchafalaya River Diversion (150,000 cfs)	03a.DI.05	★	★	★	★	★		
Bayou Lafourche Diversion (1,000 cfs)	03a.DI.01	★	★	★				
Central Wetlands Diversion (5,000 cfs)	001.DI.18	★	★	★				
East Maurepas Diversion (25,000 cfs)	001.DI.22	★	★	★				
Fort St Phillip Diversion (5,000 cfs max capacity)	001.DI.06	★	★	★				
Increase Atchafalaya Flow to Eastern Terrebonne	03b.DI.04	★	★	★	★	★		
Lower Barataria Diversion (50,000 cfs)	002.DI.15	★	★	★				
Lower Breton Diversion (50,000 cfs)	001.DI.02	★	★	★				
Mid-Barataria Diversion (60,000 cfs)	002.DI.03		★	★				
Mid-Barataria Diversion (250,000 cfs)	002.DI.04	★	★	★				
Mid-Breton Sound Diversion (5,000 cfs)	001.DI.23	★	★	★				
Upper Breton Diversion (5,000 cfs)	001.DI.14			★				
Upper Breton Diversion (250,000 cfs)	001.DI.17	★	★	★				
Wax Lake Delta Reallocation	03b.DI.05						★	★
West Maurepas Diversion (5,000 cfs)	001.DI.29	★	★	★				
West Pointe a la Hache Diversion (5,000 cfs)	002.DI.06	★	★	★				

Implementation period: ★ 2012–2031   ★ 2032–2051

RAND TR1266-4.16

Table 4.4
Frequency of Sediment Diversion Project Inclusion for Alternatives with Different Decision-Criterion Constraints (%)

Project Name	Shrimp (3)	Saltwater Fisheries (3)	Use of Natural Processes (5)	Support of Navigation (6)	O&M (5)	Critical Landforms (8)
Fort St. Phillip Diversion (5,000 cfs max capacity)	100	100	100	33	100	60
Hermitage Diversion (5,000 cfs)	100	0	100	33	100	60
Lower Breton Diversion (50,000 cfs)	100	100	75	33	100	60
Mid-Barataria Diversion (5,000 cfs)	100	100	0	0	0	0
Mid-Breton Sound Diversion (5,000 cfs)	100	100	100	33	100	60
Bayou Lafourche Diversion (1,000 cfs)	67	100	100	33	100	60
Lower Barataria Diversion (250,000 cfs)	67	0	75	17	80	40
Upper Breton Diversion (50,000 cfs)	67	0	100	17	100	60
West Pointe a la Hache Diversion (5,000 cfs)	67	67	100	33	100	60
Atchafalaya River Diversion (150,000 cfs)	33	33	100	67	100	100
Lower Barataria Diversion (5,000 cfs)	33	0	0	17	20	20
Upper Breton Diversion (5,000 cfs)	33	100	0	17	0	0
West Pointe a la Hache Diversion (250,000 cfs)	33	0	0	0	0	0
Atchafalaya River Diversion (20,000 cfs)	0	0	0	0	0	0
Bayou Lafourche Diversion (5,000 cfs)	0	0	0	0	0	0
Benneys Bay Diversion (20,000 cfs)	0	0	25	0	0	0
Bonnet Carre Diversion (5,000 cfs)	0	100	75	0	0	0
Central Wetlands Diversion (5,000 cfs)	0	0	100	33	40	40
Central Wetlands Diversion (50,000 cfs)	0	0	0	0	0	0
East Maurepas Diversion (25,000 cfs)	0	0	50	0	0	0
East Maurepas Diversion (5,000 cfs)	0	0	0	0	0	0
Hahnville Diversion (5,000 cfs)	0	0	0	0	0	0
Hermitage Diversion (250,000 cfs Seasonally Operated)	0	100	0	0	0	0

Table 4.4—Continued

Project Name	Shrimp (3)	Saltwater Fisheries (3)	Use of Natural Processes (5)	Support of Navigation (6)	O&M (5)	Critical Landforms (8)
Increase Atchafalaya Flow to Eastern Terrebonne	0	33	100	67	80	100
Lower Barataria Diversion (50,000 cfs)	0	0	0	0	0	0
Lower Breton Diversion (250,000 cfs)	0	0	0	0	0	0
Lower Breton Diversion (5,000 cfs)	0	0	0	0	0	0
Mid-Barataria Diversion (50,000 cfs)	0	0	100	33	100	60
Mid-Barataria Diversion (250,000 cfs)	0	0	0	0	0	0
Mid-Breton Sound Diversion (50,000 cfs)	0	0	0	0	0	0
Northwest Barataria Diversion (5,000 cfs)	0	0	75	0	0	0
Pontchartrain-Barataria Multi-Diversion Plan	0	0	0	0	0	40
Spanish Pass Diversion (7,000 cfs)	0	0	25	0	0	0
Third Delta Diversion (West Fork)	0	0	0	0	0	0
Upper Breton Diversion (250,000 cfs)	0	0	0	0	0	0
Violet, Davis Pond, and Bayou Lafourche Diversions (100,000 cfs)	0	0	0	0	0	0
Wax Lake Delta Reallocation	0	0	0	33	0	0
West Maurepas Diversion (5,000 cfs)	0	0	50	33	20	60
West Pointe a la Hache Diversion (50,000 cfs)	0	0	0	0	0	0

NOTE: The number of alternatives evaluated for each decision-criteria sensitivity analysis is reported in parentheses. The projects are sorted by frequency of inclusion across the alternatives varying the shrimp decision-criterion constraint.

Table 4.5
Constrained Alternatives Developed for the Master Plan

Constrained Alternative Name	Decision Criterion	Threshold Score (see Table 4.3 for definition of values)
Max Risk Reduction/Natural Processes (High)	Use of natural processes	≥–1.6
Max Risk Reduction/Natural Processes (Moderate)	Use of natural processes	≥–2.4
Max Risk Reduction/Navigation	Support of navigation	Exclude if < 0
Max Risk Reduction/Strategic Assets	Flood protection of strategic assets	≥0.174
Max Land/Sustainability	Sustainability	≥0.14
Max Land/Natural Processes (High)	Use of natural processes	≥28
Max Land/Natural Processes (Moderate)	Use of natural processes	≥24
Max Land/Navigation (High)	Support of navigation	Excluded projects with scores < –0.2
Max Land/Navigation (Moderate)	Support of navigation	Excluded projects with scores < –0.4
Max Land/Critical Landforms (High)	Critical landforms	≥0.5
Max Land/Critical Landforms (Moderate)	Critical landforms	≥0.4

provides a table that describes how each preliminary alternative is specified, as well as CPRA's rationale for including it (Table A-1).

Define the Draft Master Plan

CPRA used the Planning Tool analysis to define a single alternative for the January 2012 draft of the Master Plan (CPRA, 2012a). CPRA first reviewed the projects and outcomes for the different expert-adjusted alternatives and then selected a single alternative.

Review Projects and Outcomes for Different Alternatives

CPRA evaluated the constrained and expert-adjusted alternatives by looking at key outcomes, as calculated by the Planning Tool. Tables 4.6 and 4.7 report decision-criterion scores for each of the alternatives and the draft alternative (described in the next section). The alternative names describe the constraint or expert-adjusted decisions and environmental scenario (in some cases) under which the alternative was formulated.

Define the Final Master Plan

The draft 2012 coastal Master Plan was released on January 12, 2012, for public review and comment. CPRA held three all-day public meetings and more than 50 meetings with community groups, parish officials, legislators, and stakeholder groups. Thousands of comments were

Table 4.6
Risk-Reduction Decision-Criterion Scores for Expert-Adjusted Alternatives

Alternative	Long-Term Reduction in EAD (%)	Near-Term Reduction in EAD (%)	Use of Natural Processes, Risk Reduction (score)	Flood Protection of Historic Properties (%)	Flood Protection of Strategic Assets (%)	Risk (EAD) in Impoverished Areas ($ millions)
Draft Alternative: Modified Max Risk Reduction (moderate scenario)	70	40	−3	10.40	17.30	424
Max Risk Reduction (less optimistic scenario)	81	58	−4.1	16.30	11.20	373
Max Risk Reduction (moderate scenario)	82	59	−4.2	17.20	17.30	358
Max Risk Reduction/ Natural Processes (moderate scenario)	70	39	−2.4	7.80	11.20	423
Max Risk Reduction/ Natural Processes	67	38	−1.4	5.90	2.80	462
Max Risk Reduction/ Navigation	67	38	−1.8	5.90	2.80	448
Max Risk Reduction/No Lake Charles Levees	69	39	−2.6	10.40	17.30	436
Max Risk Reduction/ No Lake Pontchartrain Barrier	70	40	−3.9	10.40	17.30	431
Max Risk Reduction/ Nonstructural Focused	67	38	−1.8	7.60	2.80	488
Max Risk Reduction/ Strategic Assets	69	40	−3.9	10.50	17.90	465

received and reviewed and, where possible, incorporated into the final Master Plan. In addition, some of the underlying information on the individual projects was updated for accuracy. This section describes the four key steps taken to revise the draft Master Plan and define the final Master Plan:

Table 4.7
Restoration Decision-Criterion Scores for Expert-Adjusted Alternatives

Alternative	Land Area (% improvement to current conditions)	Persistence of Land (sq. mi. per decade)	Land Change in Past Decade	Use of Natural Processes, Restoration (score)	Percentage of O&M Expenditures
Draft Alternative: Modified Max Land (less optimistic scenario)	70	60.6	−175	16.4	7
Max Land (less optimistic scenario)	80	80	−124.7	20.8	8
Max Land (moderate scenario)	87	84.6	−112.9	20.1	6
Max Land/Channel Realignment	62	64.9	−163.8	15.8	10
Max Land/Critical Landforms (High)	70	85.6	−110.3	15.7	5
Max Land/Critical Landforms (Moderate)	86	84.5	−113.1	19.6	6
Max Land/Multiple Small Diversions	60	37.4	−235.1	13.9	6
Max Land/Natural Processes (High)	76	73.9	−140.6	28	9
Max Land/Natural Processes (Medium)	87	85.6	−110.1	24	7
Max Land/Navigation (Low)	82	80.8	−122.7	19.3	6
Max Land/Navigation (Medium)	87	85.7	−110	20.9	6
Max Land/No Diversions	43	−2.7	−338.9	12.5	4
Max Land/Sustainability	78	108	−52.3	20.2	7

CPRA Decisions

CPRA reviewed the constrained and expert-adjusted alternatives and their outcomes and developed a final specification to develop the plan. For risk reduction, the Master Plan is essentially a modified version of the *Max Risk Reduction* alternative with the moderate scenario. For restoration projects, the plan is a modified version of the *Max Land* alternative under the less optimistic scenario.

- Revise project data.
- Evaluate public comments.
- Revise the draft alternative for the final Master Plan.
- Review Master Plan projects and outcomes.

Revise Project Data

CPRA revised some information on the individual projects after the development of the draft Master Plan. For most projects, CPRA modified cost estimates between the draft plan and the final plan. For a limited number of projects, CPRA also modified the sources of sediment for the project, the project's description, or the project's modeled outcomes. CPRA used the Planning Tool to reevaluate its estimates of the effects of the Master Plan using the updated information. Differences in results were generally minimal.

Evaluate Public Comments

CPRA received thousands of comments during the public comment period. The comments were reviewed and categorized and are presented in the Master Plan (CPRA, 2012c, Appendix G3). Project-specific comments were used to develop adjustments to the Master Plan, which were used to revise some of the alternatives using the Planning Tool. CPRA considered project modifications where they resulted in minor or insignificant reductions in the plan outcomes. Unacceptable project modifications were instances in which a proposed change would result in large reductions in the Master Plan outcomes.

Revise the Draft Alternative for the Final Master Plan

CPRA used the Planning Tool to provide an evaluation of potential adjustments to the Master Plan as suggested by elected officials, stakeholders, and citizens through public comments. The purpose of this process was to understand the implications of making these adjustments on the plan's overall ability to meet the objectives of the Master Plan. Based on this analysis, the Master Plan was modified. For example, the final Master Plan included 15 of the 16 structural risk-reduction projects included in the draft and two additional risk-reduction projects, for a total of 17. Of the Master Plan's 93 restoration projects, all but 14 of them (or 85 percent) were carried over from the draft Master Plan.

One important change was a shift from a strict even split in funding between risk-reduction and restoration projects. CPRA's adjustments to the draft plan increased expenditures on restoration projects and decreased expenditures on risk-reduction projects, as shown in the next section.

Review Master Plan Projects and Outcomes

The Master Plan report (CPRA, 2012c) describes the final Master Plan alternative in significant detail. This section highlights some key features of the final alternative in terms of the types of projects included and the estimated outcomes for risk reduction and land area, as calculated by the Planning Tool. An assessment of Master Plan outcomes using the predictive models was performed after the Master Plan was published and is described in "Post–Master Plan Analysis, later in this chapter.

The total cost estimate of the final Master Plan is $52.1 billion (2010 dollars) over the next 50 years, and the funding split between risk-reduction and restoration projects is 41 percent risk reduction and 59 percent restoration. The differences from the draft ($50 billion total with a 50/50 split between risk reduction and restoration project funding) are due to the adjustments in the specific projects included and excluded.

Figure 4.17 shows how Master Plan funding is to be allocated across different project types and the number of projects for each type. Notably, about 20 percent of the total funding ($10.9 billion) is to be allocated to nonstructural risk-reduction projects coast-wide and $4 billion of funding is allocated to 11 different sediment-diversion projects.

The Planning Tool estimates that implementation of the Master Plan would dramatically decrease coast-wide flood risk from a currently estimated annual level of $2.4 billion on average today to between $2.4 billion and $5.5 billion in year 50 with the full implementation of the Master Plan. Without the Master Plan in place, EAD could exceed $23 billion under the less optimistic scenario. See Figure 4.18.

Using the Planning Tool's additive assumption of individual project effects, Figures 4.19 and 4.20 show how land area would change with and without the implementation of the Master Plan under the moderate and less optimistic scenarios, respectively. For the moderate scenario, coast-wide land area is stabilized by 2040 and begins to increase afterward. For the less optimistic scenario, coast-wide land area never stabilizes. Even with complete implementation of the Master Plan under the less optimistic scenario, land loss would still be severe and could be even greater than under FWOA conditions under the moderate scenario. This result suggests that it will be critical to adapt the Master Plan in the future if sea level rises and other key conditions are less favorable than those in the moderate scenario.

Figure 4.17
Master Plan Funding, by Project Type (millions of 2010 dollars)

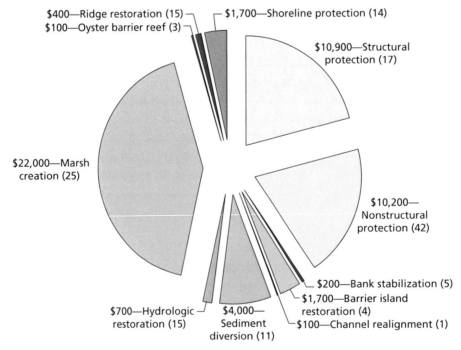

NOTE: The numbers in parentheses indicate the number of projects of each type included in the Master Plan. Funding is rounded to the nearest $100 million.
RAND TR1266-4.17

Figure 4.18
Coast-Wide Flood Risk for Current Conditions, Year 50 Without the Master Plan, and Year 50 with the Master Plan for the Moderate and Less Optimistic Scenarios

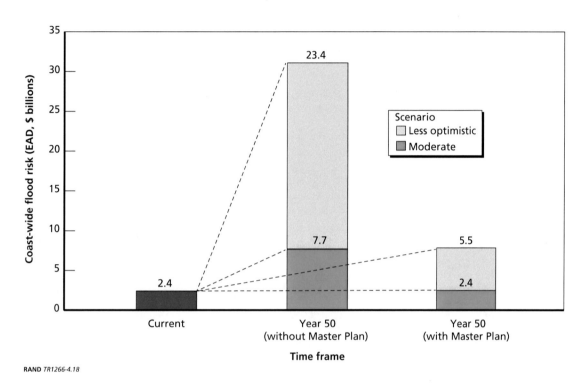

RAND TR1266-4.18

Figure 4.19
Change in Land Area With and Without the Master Plan for the Moderate Scenario

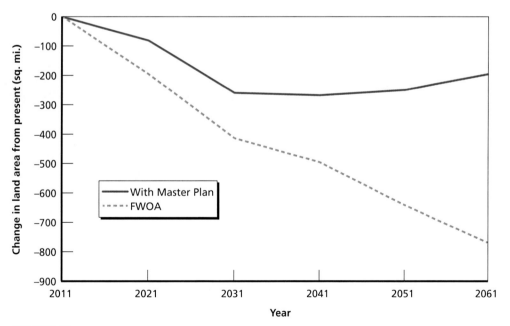

RAND TR1266-4.19

Figure 4.20
Change in Land Area With and Without the Master Plan for the Less Optimistic Scenario

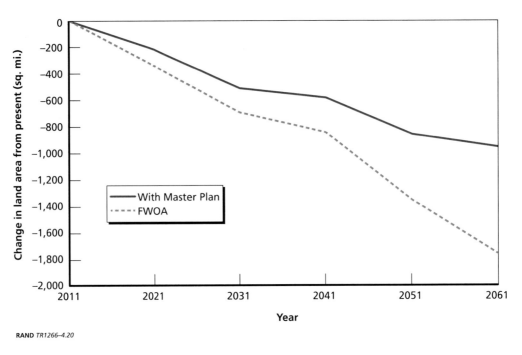

RAND *TR1266-4.20*

Post–Master Plan Analysis

CPRA initiated an integrated evaluation of the Master Plan alternative using the predictive models after the Master Plan was finalized.[7] This integrated analysis modeled the draft and final Master Plan alternatives as a single, multistage project to capture project synergies and conflicts. The integrated analysis also used predictive models and input data that were revised subsequent to the formulation of the Master Plan. As a result, the FWOA estimates for the integrated analysis (presented in this section) are different from those calculated in the Master Plan analysis.

Preliminary results shown in this section compare coast-wide risk, land, and ecosystem-service metric outcomes from the integrated analysis to the Planning Tool estimates. In general, the estimates of the key decision drivers—future risk and land area—that were made using the additive assumption within the Planning Tool were quite close to the results of the integrated analysis.

At the coast-wide scale, the Planning Tool's assumption of additive project risk-reduction effects provided a good approximation for the aggregate risk level as computed directly using the predictive models. Figure 4.21 shows coast-wide EAD in 2061 under FWOA conditions estimated for the Planning Tool (light gray bars) and for the integrated analysis (dark gray bars) and under the future with the Master Plan for the Planning Tool estimate (light purple) and the integrated analysis estimate (dark purple) under the two environmental scenarios. The flood risk estimates with the Master Plan are very similar between the Planning Tool and predictive models estimates—$2.44 billion versus $2.77 billion for the moderate scenario, and

---

[7] The final Master Plan describes some preliminary results from the analysis (CPRA, 2012a). The full analysis, however, was conducted after the Master Plan was published.

$5.49 billion versus $4.85 billion for the less optimistic scenario. Under the less optimistic scenario, there is a measurable difference in FWOA flood risk due to changes in baseline geomorphology and surge estimates.

At a more local level, differences between the Master Plan 2061 damage estimated by the Planning Tool and that estimated by the predictive models are more significant for some areas. Figure 4.22 provides results for this comparison for three communities with the largest FWOA risk: Houma, Greater New Orleans, and Slidell. For Houma, the integrated analysis shows a modest amount of damage under a future with the Master Plan ($254 million per year), whereas the Planning Tool estimates significantly less damage ($154 million per year). In contrast, the integrated analysis shows significantly less damage in Greater New Orleans both for the future without action and for the future with the Master Plan ($73 million versus $418 million under a future with the Master Plan). Lastly, Slidell shows the reverse, with higher damage estimated under the integrated analysis than by the Planning Tool ($498 million versus $274 million under a future with the Master Plan). For Greater New Orleans and Slidell, much of the difference between estimates for a future with the Master Plan is due to the changes in the FWOA baseline between the analysis for the Planning Tool and the integrated analysis.

Figure 4.21
Comparison of Coast-Wide Expected Annual Damage (billions of 2010 dollars) in 2061 Under Future-Without-Action Conditions and with Master Plan Estimates Using the Planning Tool and the Integrated Analysis for Two Environmental Scenarios

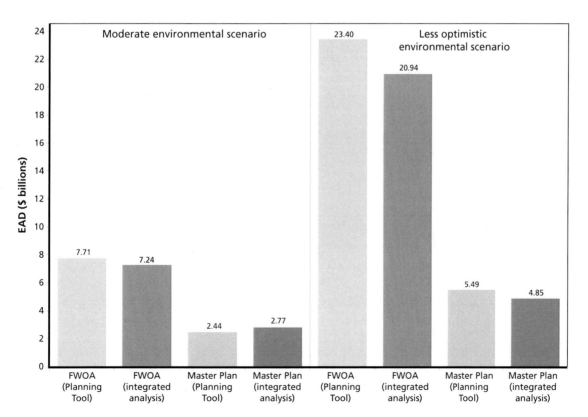

Figure 4.22
Comparison of Expected Annual Damage (millions of 2010 dollars) in 2061 for Houma, Greater New Orleans, and Slidell Under Future-Without-Action and with Master Plan Conditions Using the Planning Tool and the Integrated Analysis for the Moderate Scenario

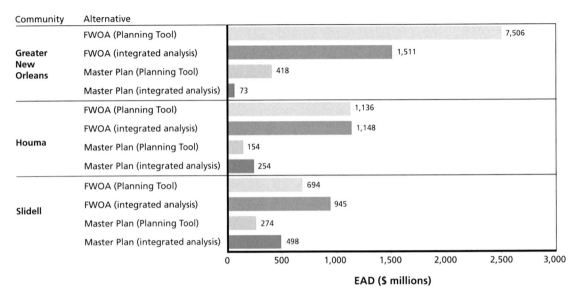

NOTE: Greater New Orleans consists of the following communities: Algiers, Arabi/Chalmette/Meraux, Avondale/Waggaman, Belle Chasse, Destrahan/New Sarpy/Norco, Metarie/Kenner, New Orleans, New Orleans East, Orleans Parish, Poydras/Violet, Saint Rose, Westbank Jefferson Parish. Houma consists of the following communities: Houma, Bayou Blue, Mathews/Lockport, and Raceland.
RAND TR1266–4.22

The integrated analysis of future land area shows more-significant differences than the estimates from the Planning Tool. These differences reflect two key factors. First, the timing of project implementation for the integrated analysis was significantly different from that used by the Planning Tool. Specifically, the integrated analysis assumed that all projects identified for implementation at year 20 (2031) and year 40 (2051) by the Planning Tool would be implemented in year 25. Second, the integrated analysis is better able to accurately reflect synergies between sediment-diversion projects and marsh-creation projects.

Figure 4.23 shows the comparison of land over time for the Planning Tool and integrated analyses. There are a few key differences. In the first 20 years, the integrated analysis estimates smaller reductions in land area than the Planning Tool. This is likely due to the positive and reinforcing interactions of the restoration projects—the effects of several projects together are greater than the sum of the individual project effects. By 2041, the integrated analysis shows a dramatic increase in land. This reflects the sequencing of all projects that were to be implemented in 2031 or 2051 to all be implemented in year 25 (2036). Many of these projects were marsh-creation projects that were specified to be implemented in later years to offset increasing sea-level rise and subsidence. Notably, all these differences across methods appear to cancel one another out by 2061; the final estimate of land is about the same for both approaches. One important implication of this analysis is that deferring some land building to later years may be essential to ensure net land building in the later time periods.

Figure 4.24 shows similar results for the less optimistic scenario. In this case, 2061 results for the integrated analysis show slightly more loss in land than the Planning Tool estimate.

Figure 4.23
Change in Land Area over Time with the Master Plan for the Moderate Scenario as Estimated by the Planning Tool and the Integrated Analysis

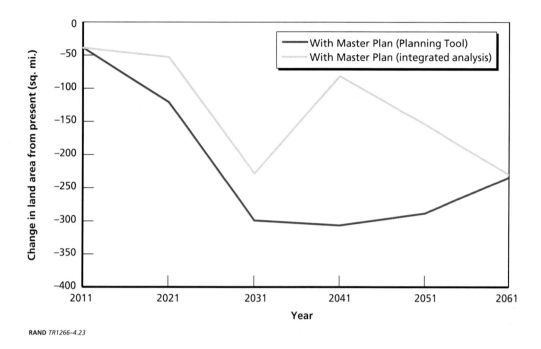

Figure 4.24
Change in Land Area over Time with the Master Plan for the Less Optimistic Scenario as Estimated by the Planning Tool and the Integrated Analysis

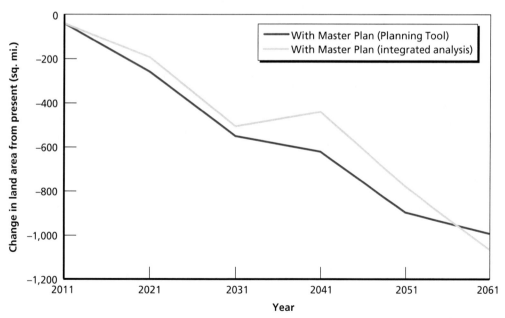

Lastly, the Planning Tool team compared for each coast-wide ecosystem-service metric the outcomes based on the Planning Tool and integrated analysis of the Master Plan. In general, the differences are significant for many metrics and suggest either that (1) the additive assumption for project effects on metrics may not estimate the combined effects of projects on the ecosystem-service metrics well or (2) the models predicting ecosystem services are unstable. The analysis performed to date cannot resolve the relative importance of the different causes of these discrepancies.

Figure 4.25 shows the ratio between the ecosystem-service metric calculated using the integrated-analysis results and the Planning Tool in 2061 for the moderate scenario. A number greater than 100 percent indicates that the predictive models estimate higher levels of the ecosystem services. For a few metrics, there is very little difference between the two methods: carbon sequestration, freshwater availability, storm-surge attenuation. Others, such as crawfish, freshwater fisheries, and other coastal wildlife, differ more than 20 percent.

The comparison of outcomes estimated by the Planning Tool (based on individual project effect estimates) with those from the integrated analysis generally reinforces the future expectations of the Master Plan alternative as described in the Master Plan (CPRA, 2012c). This is particularly the case for the two decision drivers—flood risk and coast-wide land area. For these outcomes, the coast-wide, year 50 (i.e., 2061) results vary only slightly between the Planning Tool and integrated analyses (see Figures 4.21, 4.23, and 4.24). The ecosystem-service metric outcomes, in contrast, diverge more significantly between the two approaches. Although estimates of how ecosystem services would change under different alternatives were made using the Planning Tool, they were not key drivers of the alternative-selection process. Additional analysis of the ecosystem-service outcomes are needed to better understand the sources of discrepancies between the results from the Planning Tool and integrated analysis.

Figure 4.25
Ratio of Coast-Wide Ecosystem-Service Metric Outcome for Each Ecosystem-Service Metric in Year 50 for the Moderate Scenario

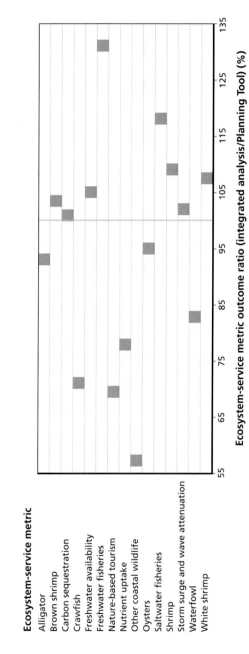

RAND TR1266-4.25

CHAPTER FIVE

Conclusions

The Planning Tool played a critical role in the development of CPRA's Master Plan. It provided a structured, analytic framework for comparing different risk-reduction and restoration projects, formulating many different alternatives, each representing one possible comprehensive approach to solving the coast risk and land-loss problems, and providing information to support the deliberation needed to formulate a single 50-year plan. The resulting 50-year Master Plan received strong public support and passed the Louisiana legislature unanimously in May 2012.

To accommodate the data and modeling available, the Planning Tool by necessity included some significant simplifying assumptions to provide useful quantitative information to the planning process. Most significant were the assumptions that individual project effects are additive, that restoration projects affect only land and ecosystem-service outcomes, and that risk-reduction projects affect only future flood risk. Preliminary analysis of the Master Plan as an integrated set of projects suggests that these assumptions were reasonable for the planning phase but will need more-careful examination as the Master Plan is implemented. In some cases, the combined effect of projects may have consequences not captured by the Planning Tool. In such cases, modifications to the plan may be necessary.

The development of the Master Plan is just the first step of a long 50-year effort to attain sustainability along Louisiana's coast. Important next steps include securing long-term funding, refining the near-term implementation strategy, and setting up a framework for adapting the plan over time. Securing funding will depend, in part, on making the case to decision-makers at the federal level that investment in Louisiana's Master Plan will yield important national benefits. The quantitative and analytic basis of the Master Plan provides technical credibility lacking in prior planning efforts, but additional work needs to be done to better quantify the benefits of implementing the Master Plan to justify public expenditures.

CPRA is now beginning work on refining in greater detail how the first phase of the Master Plan will be implemented. This work provides an opportunity to explore in greater detail the synergies and conflicts among different projects specified for implementation on a basin-by-basin level. This work will also need to consider the nature of early funding increments to maximize near-term progress by selecting the most-effective projects that can be funded given the funding source.

CPRA also made significant strides toward addressing uncertainty in its planning process through the development of two scenarios and evaluation of projects and alternatives under each. As described in "Assess Performance Under Uncertainty" in Chapter Four, CPRA chose in some cases to select projects that perform slightly less well under the moderate scenario but perform much better under the less optimistic scenario. The final analysis shows, however, that the future success of the Master Plan, as it is currently specified, is contingent in part on facing conditions that are moderate and not consistent with the less optimistic scenario. As the Master Plan moves into its implementation phase, more-significant analysis of its performance under different scenarios and potential adjustments will be helpful to ensure that the sequence of projects ultimately implemented is indeed robust to a wide range of futures. For example, if the rate of sea-level rise follows a sharper increasing trend as captured by the less optimistic scenario, the Master Plan may need to shift to the implementation of projects that were shown to be higher performers under that scenario.

Finally, the Planning Tool with some modification could be used in the analysis of other large, multiproject or multicomponent infrastructure investment challenges. These investments could relate to water supply, surface transportation, energy systems, and public housing. A modified version of the Planning Tool could support the evaluation and display of trade-offs among objectives while interacting directly with stakeholders and decisionmakers. As the experience applying the Planning Tool to the Louisiana Master Plan process showed, seeing how results change when assumptions are relaxed, constraints are changed, or projects are inserted or removed can contribute immeasurably to the transparency and credibility of a planning process.

Expert-Adjusted Alternatives

Table A.1 describes 23 different expert-adjusted alternatives, the projects included and excluded, and the motivation for developing the alternative. Each of these alternatives was specified by CPRA in consultation with CPRA management and the Master Plan stakeholders.

Table A.1
Projects Included and Excluded for Expert-Adjusted Alternatives

Name of Expert-Adjusted Alternative	Projects Excluded	Projects Included	Timing	Motivation for Examining Alternative
Max Risk Reduction Modified (moderate scenario)	Pontchartrain Barrier, SWLA GIWW	Iberia/Vermilion Upland Levee	Prevent Greater New Orleans from starting in the first time period	Moves closer to a set of projects considered feasible for implementation; Iberia/Vermilion Upland Levee protects important strategic assets
Max Risk Reduction Modified (less optimistic scenario)	Pontchartrain Barrier, SWLA GIWW	Iberia/Vermilion Upland Levee	Prevent Greater New Orleans from starting in the first time period	Moves closer to a set of projects considered feasible for implementation Iberia/Vermilion Upland Levee protects important strategic assets
Max Risk Reduction/ No Lake Charles Levees or SWLA GIWW	Lake Charles Levees, SWLA GIWW			Explores EAD reductions in Lake Charles if all structural options were removed
Max Risk Reduction/ No Morganza to the Gulf	Morganza to the Gulf			Explored because of stakeholder interest; when Morganza to the Gulf was removed, Donaldsonville to the Gulf was then selected
Max Risk Reduction/ No Pontchartrain Barrier	Both versions of the Lake Pontchartrain Barrier			Explored because of stakeholder interest and the barriers' potential negative impacts on Mississippi
Max Risk Reduction/ Nonstructural Focused	Cross-Basin Barrier Projects			Explored to provide better comparison of structural and nonstructural options

Table A.1—Continued

Name of Expert-Adjusted Alternative	Projects Excluded	Projects Included	Timing	Motivation for Examining Alternative
Max Land Modified (moderate scenario)	GIWW Lock West of Calcasieu, Pass a Loutre Marsh Creation, Eugene Island to Pointe Au Fer Island Oyster Barrier, Little Pecan Bayou, Third Delta	HNC Lock; specific diversions		A strategic combination to help streamline alternatives by avoiding duplicative projects. HNC Lock is a key element of Morganza to the Gulf, a project always selected for EAD reduction. There was a consensus among stakeholders that Third Delta should not be included.
Max Risk Reduction Modified (moderate scenario)	Pontchartrain Barrier, SWLA GIWW	Iberia/Vermilion Upland Levee	Prevent Greater New Orleans High Level Plan from starting in the first time period	Moves closer to a set of projects considered feasible for implementation; Iberia/Vermilion Upland Levee protects important strategic assets
Max Risk Reduction Modified (less optimistic scenario)	Pontchartrain Barrier, SWLA GIWW	Iberia/Vermilion Upland Levee	Prevent Greater New Orleans High Level Plan from starting in the first time period	Moves closer to a set of projects considered feasible for implementation; Iberia/Vermilion Upland Levee protects important strategic assets
Max Risk Reduction/ SWLA Levees	Lake Charles Levees, SWLA GIWW, Pontchartrain Barrier			Explores EAD reductions in Lake Charles and SWLA if all structural options were removed
Max Risk Reduction/ No Morganza to the Gulf	Morganza to the Gulf, Pontchartrain Barrier			Explored because of environmental concerns
Max Risk Reduction/ No Pontchartrain Barrier	Both versions of the Lake Pontchartrain Barrier			Explored because of stakeholder interest and the barriers potential negative impacts on Mississippi
Max Risk Reduction/ Nonstructural Focused	Cross-Basin Barrier Projects, Pontchartrain Barrier			Explored to provide better comparison of structural and nonstructural options
Max Land Modified (moderate scenario)	GIWW Lock West of Calcasieu, Pass a Loutre Marsh Creation, Eugene Island to Pointe Au Fer Island Oyster Barrier, Little Pecan Bayou, Third Delta	HNC Lock; specific diversions		A strategic combination to help streamline duplicative projects; HNC Lock is a key element of Morganza to the Gulf, a project always selected for EAD reduction

Table A.1—Continued

Name of Expert-Adjusted Alternative	Projects Excluded	Projects Included	Timing	Motivation for Examining Alternative
Max Land Modified (less optimistic scenario)	GIWW Lock West of Calcasieu, Pass a Loutre Marsh Creation, Eugene Island to Pointe Au Fer Island Oyster Barrier, Little Pecan Bayou, Third Delta	HNC Lock; specific diversions		A strategic combination to help streamline duplicative projects; HNC Lock is a key element of Morganza to the Gulf, a project always selected for EAD reduction
Max Land Hybrid	GIWW Lock West of Calcasieu, Pass a Loutre Marsh Creation, Eugene Island to Pointe Au Fer Island Oyster Barrier, Little Pecan Bayou, Third Delta	HNC Lock; specific diversions		A strategic combination of projects that are good for moderate and less optimistic scenarios and helps streamline duplicative projects; HNC Lock is a key element of Morganza to the Gulf, a project always selected for EAD reduction
Max Land/No Diversions	All diversions and channel realignments			Explored because of stakeholder interest
Max Land/Small Diversions I	Pontchartrain-Barataria Multi-Diversion Plan			Explored inclusion of multiple small diversions in place of other diversion with which they are mutually exclusive because of stakeholder interest
Max Land/Channel Realignment		Channel Realignment 80/20		Explored role of channel realignments because they were never chosen in other alternatives
Max Land/No Locks	HNC Lock, Salinity Control on Calcasieu Ship Channel			Explored to understand impact of locks
Max Land/With Locks		HNC Lock, Salinity Control on Calcasieu Ship Channel		Explored to understand impact of locks
Max Land/ No Atchafalaya Diversions	All Atchafalaya diversions			Explored to understand what happens without these diversions because they were always selected
Max Land/Limited Hydrological Restoration	Seven Hydrologic Restoration Projects			Explored because of stakeholder interest

NOTE: SWLA = southwest Louisiana. HNC = Houma Navigation Canal.

Glossary

additive assumption. A Planning Tool assumption in which the combined effects of individual projects on the coast are estimated by adding estimates of individual project effects.

alternative. A group of individual risk-reduction and restoration projects and the time period in which they are to be implemented.

constraint. A limitation or restriction on how projects can be grouped together to make up an alternative—the Planning Tool implements a set of constraints mathematically.

cost-effectiveness. A performance metric derived by dividing a project's specific effect (e.g., on land area) by the cost of the project.

cultural heritage. A Master Plan decision criterion designed to capture changes in the ability of Louisiana's unique culture to thrive in the coming decade.

decision criterion. A specific outcome calculated by a model or the Planning Tool that is relevant to the formulation of alternatives; the Planning Tool can require that an alternative's score for one or more decision criteria exceed a particular level.

decision driver. One of two primary factors used to compare projects and develop alternatives—coast-wide flood risk and coast-wide land area.

decision variable. An indicator in the Planning Tool objective function of whether a particular project is started during a particular implementation period for a given alternative.

draft Master Plan. The January 2012 draft of *Louisiana's Comprehensive Master Plan for a Sustainable Coast* (CPRA, 2012a).

ecosystem-service metric. A quantitative measure that approximates the level of an ecosystem service, such as shrimp habitat, freshwater availability, and storm-surge attenuation.

expected annual damage. The monetary damage that would occur on average as a result of flooding from category 3 or greater storms in any given year, if a particular region were subjected to the same specific conditions and probability distribution of flood depths over many years.

expert-adjusted alternative. An alternative formulated by the Planning Tool that is modified by including or excluding specific projects.

final Master Plan. The spring 2012 final of *Louisiana's Comprehensive Master Plan for a Sustainable Coast* (CPRA, 2012c).

formulating alternatives. The use of the Planning Tool to develop groups of projects (or alternatives) that differ in their emphasis on different Master Plan objectives or by the constraints applied.

funding split. The shares of funding that are allocated to risk-reduction and restoration projects.

future without action. The future condition in which no additional risk-reduction or restoration projects are implemented.

integrated analysis of Master Plan. The analysis of the Master Plan alternative by the predictive models in which all projects are modeled concurrently to capture project synergies and conflicts.

long term. The later 30 years of the planning time horizon (2032–2061).

mixed-integer program. A type of optimization algorithm that maximizes an objective function that includes both integer and noninteger variables; the Planning Tool uses a mixed-integer program to formulate alternatives.

multicriterion decision analysis. An analytic approach for ranking different decisions based on their effects on more than a single criterion.

near term. The first 20 years of the planning time horizon (2012–2031).

nonstructural protection. Flood risk reduction achieved through the elevation, floodproofing, or removal of buildings; nonstructural protection projects in the Master Plan are defined for individual communities.

objective function. The mathematical statement that is maximized by the Planning Tool when formulating alternatives; the Planning Tool's objective function includes the weighted sum of near-term and long-term risk reduction and land building.

outcome. Coastal conditions that are predicted by the predictive models.

Planning Tool analysis of the Master Plan. The estimates of the Master Plan's effects on the coast made by the Planning Tool using individual project effect estimates and the additive assumption.

Planning Tool results visualizer. A compilation of interactive visualizations based on Planning Tool analysis; visualizations were developed using the commercially available software package Tableau Desktop; a free viewer version of Tableau is available for Microsoft Windows–based computers.

predictive models. A set of linked computer models of the coastal system that is used to predict outcomes under future-without-action and with-project conditions.

project. A single risk-reduction and restoration intervention evaluated by the Master Plan; the Master Plan is made up of a group of risk-reduction and restoration projects.

project characteristics. Quantitative information about projects used by the predictive models and the Planning Tool.

project effects. The changes in outcomes due to the implementation of a project as estimated by the predictive models.

residual damage. The level of flood risk damage to physical assets (in dollars) that would be exceeded in a given year with a specific frequency of recurrence (for example, one in 100 years).

scenario (environmental). A set of values for uncertain factors used by the predictive models to reflect uncertainty about future coastal conditions; two environmental scenarios were evaluated by the Master Plan: moderate and less optimistic.

scenario (funding). A specification of available funding over time that is used by the Planning Tool to formulate alternatives; two primary funding scenarios were evaluated by the Master Plan: low funding ($20 billion) and high funding ($50 billion).

sediment diversion. A type of restoration project that enables the diversion of river water and sediment from the main channel to wetlands for purposes of nourishing them.

structural protection. A type of risk-reduction project that uses large, structural infrastructure, such as a levee or flood wall.

References

Brinkley, Douglas, *The Great Deluge: Hurricane Katrina, New Orleans, and the Mississippi Gulf Coast*, New York: HarperCollins, 2006.

Coastal Protection and Restoration Authority, *Louisiana's Comprehensive Master Plan for a Sustainable Coast*, draft, Baton Rouge, La., January 2012a.

———, "Stakeholder Review," 2012b. As of August 14, 2012:
http://www.coastalmasterplan.louisiana.gov/working-together/stakeholder-review/

———, *Louisiana's Comprehensive Master Plan for a Sustainable Coast*, Baton Rouge, La., May 22, 2012c. As of August 14, 2012:
http://www.coastalmasterplan.louisiana.gov/2012-master-plan/final-master-plan/

Couvillion, Brady R., John A. Barras, Gregory D. Steyer, William Sleavin, Michelle Fischer, Holly Beck, Nadine Trahan, Brad Griffin, and David Heckman, *Land Area Change in Coastal Louisiana from 1932 to 2010*, Lafayette, La.: U.S. Geological Survey Wetlands Research Center, U.S. Geological Survey Scientific Investigations Map 3164, June 1, 2011. As of August 14, 2012:
http://purl.fdlp.gov/GPO/gpo8208

CPRA—*See* Coastal Protection and Restoration Authority.

Fischbach, Jordan R., David R. Johnson, David S. Ortiz, Benjamin Bryant, Matthew Hoover, and Jordan Ostwald, *Coastal Louisiana Risk Assessment Model*, Santa Monica, Calif.: RAND Corporation, forthcoming.

Grossi, Patricia, and Robert Muir-Wood, *Flood Risk in New Orleans: Implications for Future Management and Insurability*, Newark, Calif.: Risk Management Solutions, 2006. As of April 30, 2012:
http://www.rms.com/Publications/NO_FloodRisk.pdf

Keeney, Ralph L., and Howard Raiffa, *Decisions with Multiple Objectives: Preferences and Value Tradeoffs*, New York: Cambridge University Press, 1993.

Kiker, Gregory A., Todd S. Bridges, Arun Varghese, P. T. Seager, and Igor Linkov, "Application of Multicriteria Decision Analysis in Environmental Decision Making," *Integrated Environmental Assessment and Management*, Vol. 1, No. 2, April 2005, pp. 95–108.

Knutson, Thomas R., John L. McBride, Johnny Chan, Kerry Emanuel, Greg Holland, Chris Landsea, Isaac Held, James P. Kossin, A. K. Srivastava, and Masato Sugi, "Tropical Cyclones and Climate Change," *Nature Geoscience*, Vol. 3, No. 3, 2010, pp. 157–163.

Kolker, Alexander S., Mead A. Allison, and Sultan Hameed, "An Evaluation of Subsidence Rates and Sea-Level Variability in the Northern Gulf of Mexico," *Geophysical Research Letters*, Vol. 38, No. 21, 2011, p. L21404.

Lahdelma, Risto, Pekka Salminen, and Joonas Hokkanen, "Using Multicriteria Methods in Environmental Planning and Management," *Environmental Management*, Vol. 26, No. 6, 2000, pp. 595–605.

Lempert, Robert J., Steven W. Popper, and Steven C. Bankes, *Shaping the Next One Hundred Years: New Methods for Quantitative, Long-Term Policy Analysis*, Santa Monica, Calif: RAND Corporation, MR-1626-RPC, 2003. As of August 14, 2012:
http://www.rand.org/pubs/monograph_reports/MR1626.html

Linkov, I., F. K. Satterstrom, G. Kiker, C. Batchelor, T. Brides, and E. Ferguson, "From Comparative Risk Assessment to Multi-Criteria Decision Analysis and Adaptive Management: Recent Developments and Applications," *Environment International*, Vol. 32, No. 8, December 2006, pp. 1072–1093.

National Research Council, Panel on Strategies and Methods for Climate-Related Decision Support, *Informing Decisions in a Changing Climate*, Washington, D.C.: National Academies Press, 2009.

Romero, Carlos, *Handbook of Critical Issues in Goal Programming*, New York: Pergamon Press, 1991.

Schrijver, Alexander, *Theory of Linear and Integer Programming*, New York: Wiley, 1998.

USACE—*See* U.S. Army Corps of Engineers.

U.S. Army Corps of Engineers, *Louisiana Coastal Protection and Restoration: Final Technical Report*, New Orleans, La., 2009.

U.S. Census Bureau, "5-Year Release Details," American Community Survey, last revised June 27, 2012. As of August 2, 2012:
http://www.census.gov/acs/www/data_documentation/2009_5yr_data/

Vermeer, Martin, and Stefan Rahmstorf, "Global Sea Level Linked to Global Temperature," *Proceedings of the National Academy of Sciences of the United States of America*, Vol. 106, No. 51, 2009, pp. 21527–21532.